Deciphering the
Cosmic Number

Also by Arthur I. Miller

Empire of the Stars: Obsession, Friendship, and Betrayal in the Quest for Black Holes

Einstein, Picasso: Space, Time, and the Beauty That Causes Havoc

Insights of Genius: Imagery and Creativity in Science and Art

Imagery in Scientific Thought: Creating 20th-Century Physics

Deciphering the Cosmic Number

The Strange Friendship of Wolfgang Pauli and Carl Jung

Arthur I. Miller

W·W· Norton & Company   New York   London

Copyright © 2009 by Arthur I. Miller

For information about permission to reproduce selections from this book, write to
Permissions, W. W. Norton & Company, Inc., 500 Fifth Avenue, New York, NY 10110

For information about special discounts for bulk purchases, please contact
W. W. Norton Special Sales at specialsales@wwnorton.com or 800-233-4830

Manufacturing by R. R. Donnelley, Harrisonburg
Book design by Margaret M. Wagner
Production manager: Julia Druskin

Library of Congress Cataloging-in-Publication Data

Miller, Arthur I.
    Deciphering the cosmic number : the strange friendship of Wolfgang Pauli and
Carl Jung / Arthur I. Miller.
        p. cm.
    Includes bibliographical references and index.
    ISBN 978-0-393-06532-9 (hardcover)
    1. Pauli, Wolfgang, 1900–1958. 2. Jung, C. G. (Carl Gustav), 1875–1961.
3. Numerology. 4. Symbolism of numbers. 5. Physics—Philosophy. I. Title.
QC16.P37M55 2009
154.2—dc22

                                                    2008047775

W. W. Norton & Company, Inc.
500 Fifth Avenue, New York, N. Y. 10110
www.wwnorton.com

W. W. Norton & Company Ltd.
Castle House, 75/76 Wells Street, London W1T 3QT

1 2 3 4 5 6 7 8 9 0

To Lesley

Contents

Acknowledgments

LOOKING into the story of Carl Jung and Wolfgang Pauli has taken me on a journey into ways of exploring the cosmos that transcend psychology and physics and transported me to areas to which I had never before given serious thought. A supposedly rational physicist and historian, I found myself investigating alchemy, mysticism, and the Kabbalah.

I owe an enormous debt to my friend and colleague Karl von Meyenn who opened many doors to me in my study of Pauli's life. For over thirty years Karl has worked on editing Pauli's vast correspondence, now published in eight splendid volumes. He was extremely generous in sharing unpublished insights and documents for which I am hugely appreciative.

Early in my research I had the good fortune to meet Carl Jung's grandson, Andreas Jung. He graciously received me at 228 Seestrasse, Küsnacht, near Zürich, once his grandfather's house. He gave me a guided tour, showing me Jung's small study and large library crammed with esoteric books as well as the dining room, which Jung considered the center of the house. It was in these rooms that Pauli sat as Jung's patient and then as esteemed colleague and co-worker.

Information collected in archives is indispensable for historical work. I am grateful to Anita Hollier, the archivist at CERN who oversees *La Salle Pauli* where Pauli's physics papers and personal books are stored. She patiently guided me through it as well as the magnificent CERN Document Server and Pauli Photo Archive.

Thanks to Gabriele Veneziano, chair of the Pauli Committee at CERN, for his kind assistance in my research and for many good conversations on the nature of things. I would also like to take this occasion to express my gratitude to the Pauli Committee for their kind considerations of my requests for access to archival materials.

Important archival material relating to Jung and Pauli is housed at the ETH-Bibliothek Archive. Michael Gasser, head of Archives and Private Collections, Rudolf Mumenthaler, and Yvonne Voegeli facilitated my access to this collection. A special thank-you for their splendid hospitality.

I found enlightening information on Pauli's sister, Hertha, at the Manuscripts Division at the New York Public Library, which I thank for their assistance. Thanks to Susanne Blumesberger, Ursula Gabel, Christian Gastberger, and Charles Enz, Pauli's last assistant, for informative conversations on Hertha.

Interviews with Igal Talmi, at the Weizmann Institute, Tel Aviv, and T. D. Lee at Columbia University, New York City, broadened my knowledge of Pauli the man.

Ullrich Müller-Herold and Norbert Straumann took me on several enjoyable strolls around Zürich and filled me in on the scientific milieu there during the last years of Pauli's life.

Helmut Rechenberg told me a great deal about Pauli's time in Munich as a student and his relationship with Heisenberg.

Sonu Shamdasani generously made available to me some of Jung's unpublished lectures and informed me about recent developments in Jung scholarship. Another Jung scholar, Angela Graf-Nold, helped me navigate Jung material at the ETH-Bibliothek and provided me with new Jung sources and information about his professional life in Zürich. My thanks to both.

At the May 2007 symposium on Jung and Pauli, in scenic Ascona in Switzerland, I was fortunate enough to meet a number of "Jungians" who have been extremely helpful as well as becoming friends. Special thanks to Reinhard Nesper, Harald Atmanspacher, and Suzanne Gieser. Suzanne's scholarship has been enormously helpful to me.

I appreciate informative exchanges with Finn Aaserud and Herbert van Erkelens.

In Jerusalem Josef Dan, one of the world's foremost experts on the Kabbalah, gave me valuable insights into the subject.

Thanks to Hans-Joachim Braun and Karin Reich for wonderful historical tours of the University of Hamburg, where Pauli held his first professorship.

John Barrow, Jan Munch Pederson, and Simon Singh kindly replied to questions about "strange numbers."

Thanks to Chiara Ambrosio for her assistance in gathering source materials and for chats about creativity.

Conversations with Jeremy Bernstein, Freeman J. Dyson, and T. D. Lee were valuable for my investigation of events surrounding Pauli's 1958 lecture at Columbia University, for which I am grateful.

For perceptive comments on the manuscript, I thank Mike Brady, Karl von Meyenn, and Sonu Shamdasani. I am especially grateful to Gary Steigman for his insightful and detailed critiques, which were immensely helpful.

As always, my agent and good friend Peter Tallack of The Science Factory has been a pillar of support and enthusiasm, providing sagacious advice and comments on successive drafts.

I am hugely grateful to my editor at W. W. Norton, Angela von der Lippe, for her encouragement and for her many valuable criticisms. I could not have written this book without her help. Thanks too to Erica Stern for easing me through production hurdles.

Unless indicated otherwise, all translations are mine.

My primary interest has always been in studying the creative process. The interaction between Jung and Pauli is a powerful example. To unravel the equations of the soul, they embarked on a path that led them deep into the psychology of the unconscious, which Jung called the "darkest hunting ground of our times." To tell their story I have spun a scenario based on available information. In this way I hoped to look into their minds and understand better who these men really were.

Chapters 8 and 9 explore Jung's analysis of Pauli's dreams. We cannot know exactly what transpired between them in the privacy of Jung's study. I have inferred the scenario in these chapters on the basis of the in-depth descriptions Jung made soon afterward and Pauli's biographical details.

For all this I bear full responsibility. Any errors that remain are my own.

Many thanks to those who provided me with photographs and who helped me locate them as well as their copyright owners. Every effort has been made to trace copyright holders; if any have been missed, I would appreciate them contacting me.

My wife, Lesley, as always full of good cheer and love, provided me with peace of mind and indispensable encouragement. She is also a fount of invaluable advice on how to turn out a readable book. I am indebted to her for all this and for much else. This book is dedicated to her.

Arthur I. Miller
London, 2008
www.arthurimiller.com

The no-man's land between Physics and the Psychology of the Unconscious [is] the most fascinating yet the darkest hunting ground of our times.—CARL JUNG

What is decisive for me is that I *dream* about physics as Mr. Jung (and other non-physicists) *think* about physics. Every time I have talked to Mr. Jung (about the "synchronistic" phenomenon and such), a certain spiritual fertilization takes place.—WOLFGANG PAULI

Prologue

Is THERE a number at the root of universe? Is there a primal number? Is there a number that everything in the universe hinges on, that explains everything? Many of the major discoveries in science have emerged out of mathematics—Einstein's general theory of relativity, black holes, parallel universes, string theory, and complexity theory are only a few of many examples. All of these can be expressed in equations; yet they also depict concrete aspects of the physical universe.

Could there be a single number at the root of the universe which is, as Douglas Adams has it in *The Hitchhiker's Guide to the Galaxy*, "the answer to life, the universe and everything?" Physicists, psychologists, and mystics have pondered this question. Some have proposed the number three—as in the Trinity and the three dimensions of length, breadth, and depth. Some have argued for four—after all, we have four seasons, four directions (north, south, east, and west), and four limbs. Some have been convinced that the answer might be the very weird number 137, which on the one hand very precisely describes the DNA of light and on the other is the sum of the Hebrew letters of the word "Kabbalah." This is a matter that exercised many of the great minds of the twentieth century, among them the physicist Wolfgang Pauli and the psychoanalyst Carl Jung.

Deciphering the Cosmic Number is the story of two mavericks—Pauli, the scientist who dabbled in the occult, and Jung, the psychologist who was sure that science held answers to some of the questions that tormented him. Both made enormous and lasting contributions to their fields. But in their many conversations they went much further, exploring the middle ground between their two fields and striking sparks off each other.

In 1931 Wolfgang Pauli was at the height of his scientific career. He had discovered the exclusion principle—known to this day as the Pauli exclusion principle—which explains why the structure of matter is as it is and why certain stars die as they do.

Just a year earlier, he had made the audacious suggestion that there might be an as yet undiscovered particle—an outrageous suggestion in those days. Besides the electron, proton, and light quantum, which everyone took for granted, he insisted that there had to be another particle that became known as the neutrino. Twenty-six years later Pauli's neutrino was finally discovered in the laboratory.

But while his friends and colleagues competed to win science's glittering prizes, Pauli was a different kind of character. He seemed almost indifferent to success. His scientific work was not enough to give him satisfaction and his personal life too fell deeper and deeper into chaos as he trawled the bars of Hamburg, sampling the nightlife and chasing after women.

In 1932 a prize-winning film of *Dr. Jekyll and Mr. Hyde* came out, starring Frederic March as the tormented doctor. Pauli's life too seemed to have fractured.

The solution was obvious. He turned to the world-famous psychologist Carl Jung who, as it turned out, lived not far from him just outside Zürich.

Pauli was thirty-one. Jung, his senior by twenty-six years, was firmly established and hugely famous. He was the toast of the wealthy ladies and gentlemen of European and American high society, who came to him hoping to solve their various psychological malaises.

At the time the world was still living through the aftermath of the 1929 Wall Street crash; two years earlier in Germany the Nazis had won 37 percent of the vote in a key election and Adolf Hitler was on the way to becoming chancellor; Japan had recently invaded Manchuria; and Franklin Delano Roosevelt had just been elected president of the United States. But none of this much affected Jung and his wealthy patients. They were interested in more arcane and intimate matters.

Along with Sigmund Freud, Jung had opened up the concept of the mind as something that could be studied and understood—and also healed. But the approaches of the two legendary psychoanalysts could not have been more different.

Right from the start Jung wanted to shed light on those deep recesses of the unconscious that were beyond Freud's method, which dealt only with the areas of the unconscious generated by events in one's daily life. Yet Jung was far more than just a psychologist. His interests ranged far and wide across Chinese philosophy, to alchemy and UFOs. He saw the same patterns underlying radically different ways of thinking across the world, and he was convinced that these patterns arose from the mind. He called them archetypes, essential elements of the pysche. Thus he developed the concepts of the collective unconscious and of archetypes, which are today taken for granted.

He then came up with the concept of synchronicity, which he always considered one of his most important ideas. He was sure that bonds as strong as those that linked Eastern and Western thinking could also link the apparently cold rational world of science with the supposedly irrational world of intuition and the psyche.

One area that brought all these interests together was numbers. Jung was fascinated by certain numbers—three and four—that popped up again and again in alchemy and also in religion, and in the power of numbers to predict occurrences in life, as codified in the *I Ching* (the Chinese Book of Changes). But it was not until he met Wolfgang Pauli that all this began to coalesce.

/ / /

PAULI, a kindred spirit, was also fascinated by numbers. His infatuation with numbers had begun when he was a physics student, when his mentor Arnold Sommerfeld used to extol the wonders of whole numbers with all the fervor of a kabbalist. Among them was 137.

It was Sommerfeld who discovered this extraordinary number in 1915, while trying to solve one particular puzzling feature of atoms: the "fine structure" of spectral lines, the characteristic combination of wavelengths of light emitted and absorbed by each chemical element—the fingerprint or DNA, as it were, of each wavelength of light. It was dubbed the "fine structure constant" (which in fact equals 1/137, though for convenience physicists refer to it as 137).* From the moment 137 first popped up in his equations, he and other physicists saw that its importance went far beyond the fact that it solved this one puzzle. They quickly realized that this unique "fingerprint" was the sum of certain fundamental constants of nature, specific quantities believed to be invariable throughout the universe, quantities central to relativity and the quantum theory.

But if this one number were so important, should it not be possible to deduce it from the mathematics of these theories? Disturbingly, no one could.

The fine structure constant turns out to be exquisitely tuned to allow life as we know it to exist on our planet. Perhaps it was not surprising, then, that physicists began referring to 137 as a "mystical number."

By the time Sommerfeld stumbled across 137 in 1915, whole numbers were beginning to crop up everywhere in atomic physics. Two years before, the Danish physicist Niels Bohr had worked out that the energy levels of the electrons within atoms could be expressed with whole numbers, so-called quantum numbers. He assumed that only three quantum numbers were necessary to locate an electron in the atom, just as it takes only three numbers to locate an object in space: its coordinates in the three dimensions. But then ten years later the twenty-four-year-old Pauli showed that in fact a fourth quantum number was needed. The problem was that the fourth quantum number could not be visualized.

For Pauli the problem came down to numbers: to the "difficult tran-

---

*In fact, the number Sommerfeld initially calculated for the fine structure was 0.00729. The road to how and why it was rewritten as 1/137 will unfold in later chapters.

sition from three to four." And 137 turned out to be linked with this transition.

Three hundred years earlier, a full-scale row over a very similar issue had broken out between the mystic and scientist Johannes Kepler and the Rosicrucian Robert Fludd. Kepler argued that three was the fundamental number at the core of the universe, using arguments from Christian theology and ancient mysticism. Fludd, however, argued for four on the basis of the Kabbalah, of the four limbs, the four seasons, and the four elements (earth, water, air, and fire): God's creation of the world was a transition from two to three to fourness, he asserted.

But where did 137 come in? Pauli became convinced that the number was so fundamental that it ought to be deducible from a theory of elementary particles. This quest took over his waking and sleeping life. Driven beyond endurance, he sought the help of Jung.

Jung's theory of psychology offered Pauli a way to understand the deeper meaning of the fourth quantum number and its connection with 137, one that went beyond science into the realm of mysticism, alchemy, and archetypes. Jung, for his part, saw in Pauli a treasure trove of archaic memories, as well as a great scientist who could help him put his theories on a firm footing.

THE EARLY YEARS of the twentieth century were a watershed not unlike the Renaissance. Freud's discovery of the mind as a field of study and Max Planck's discovery of the quantum nature of matter were quickly followed by Einstein's relativity theory and Bohr's theory of the atom. Then came the horrors of the First World War, which inspired a trend toward spiritualism and a return to ancient beliefs, especially in Germany. Just before the war the great German physicist Werner Heisenberg was finding solace in reading Plato. In 1927 Sommerfeld, in response to a request by a periodical for an article on astrology, wrote:

> Doesn't it strike one as a monstrous anachronism that in the twentieth century a respected periodical sees itself compelled to solicit a discussion about astrology? That wide circles of the educated or half-educated public are attracted more by astrology than astronomy?

[We] are thus evidently confronted once again with a wave of irrationality and romanticism like that which a hundred years ago spread over Europe as a reaction against the rationalism of the eighteenth century.

Yet he himself wrote ecstatically of the mystical qualities of 137.

The search for some point of contact between physics and the mind was of key interest to many physicists, including Max Born and Werner Heisenberg—two other pioneers of quantum physics—and Pauli and Bohr. As Pauli put it:

I do not believe in the possible future of mysticism in the old form. However, I do believe that the natural sciences will out of themselves bring forth a counter pole in their adherents, which connects with the old mystic elements.

All this was taking place at a time when philosophy was shifting from a positivistic approach, which excluded anything that could not be reduced to sense perceptions, to a search for a reality beyond appearances. The search for this reality became a passionate quest in the arts as well: Pablo Picasso and Wassily Kandinsky were discovering new ways to represent reality as they developed cubism and abstract expressionism; composers such as Igor Stravinsky and Arnold Schönberg were rebelling against the traditional canons of music; while writers such as James Joyce were incorporating relativity into their fiction.

PAULI told very few colleagues about his discussions with Jung. He feared their derision. Nevertheless his sessions with Jung convinced him that intuition rather than logical thought held the key to understanding the world around us. Many scientists see Pauli as the epitome of rationality and logical thinking. They assume that a scientist who worked as hard as he did, and achieved as much, must have lived strictly a life of the mind, devoted to physics. This still tends to be the image that both ordinary people and scientists themselves have of scientists.

It is important to remember Isaac Newton, who laid the foundations of modern science. For over two hundred years after his death people imagined he was a man devoid of emotions—"with his Prism and silent Face," as William Wordsworth wrote—who sat at his desk day after day working out equations.

A colleague once asked Newton what he was working on. He replied that he did physics—but only in his spare time. In the 1930s, a bundle of papers which he had kept secret came to light. These revealed that Newton had been very much a man of his time, concerned less with physics than with issues such as how big the new city of Jerusalem would have to be to receive the souls on Judgment Day, with biblical chronology and how to discern the motion of material objects relative to God. As far as he was concerned, his famous laws of motion were simply a means to work toward this end.

As the English economist John Maynard Keynes, who bought many of Newton's newly discovered papers, wrote, "Newton was not the first of the age of reason. He was the last magician."

Newton's first biographer, the nineteenth-century Scottish scientist David Brewster, was adamant that there was "no reason to suppose that Sir Isaac Newton was a believer in the doctrines of alchemy." But Newton's papers reveal just the opposite—that Newton was among the most knowledgeable alchemists of his day. We now take for granted that he should be understood as a man of his time, who lived in a world of alchemy, magic, and mysticism, like his near-contemporary, the seventeenth-century German astronomer Johannes Kepler, whom Pauli saw as an image of himself.

Scientists who have not examined Pauli's vast correspondence and writings still place him in the old Newtonian straitjacket. But Pauli was alive to the alchemical roots of science. Modern science, he believed, had come to a dead end. Perhaps the means to break through and to develop new insights was to take a radically different approach and return to science's alchemical roots.

Although a twentieth-century scientist, Pauli felt an affinity with the seventeenth century—perfectly natural to anyone who, as he did, accepted that there was, as Jung postulated, a collective unconscious.

Today a vocal minority of scientists believe in paranormal phenomena. For twenty eight years a laboratory at Princeton University tried to establish evidence for extra-sensory perception (ESP)—using card-guessing methods—as well as evidence for telekinesis, the ability of the mind to move objects. It had been privately funded to the tune of ten million dollars and closed down in 2007. Its founder, Robert G. Jahn, a pioneer in jet propulsion systems said, "it is time." He claimed to have demonstrated that test subjects "thinking high" and "thinking low" could alter a sequence of numbers flashed from a random number generator—very slightly, however, two or three flips out of ten thousand. Pauli and Jung discussed experiments of this sort. They, too, believed in powers of the mind inexplicable by the logic of physics.

The two men also discussed at great length the notion of consciousness, considered by most scientists at that time to be sheer nonsense—"off limits." Today it is a burgeoning field of research using concepts from quantum mechanics, some of which Pauli had speculated on.

MORE THAN TWENTY YEARS ago I was intrigued to discover that Pauli and Jung had co-authored a book entitled *The Interpretation of Nature and the Psyche.* I tracked it down and read it with growing fascination. I was gripped by the new aspects of both men it revealed. As a physicist I knew about Pauli and his contributions to science and of course was well aware of Jung. But the two together, the rational Pauli with the iconoclastic Jung?

I was determined to find out more about their story. Nevertheless, many years passed before I finally had the chance. I began my research in Zürich where I studied their letters, housed in the library of the very famous technological university, the ETH (*Eidgenössische Technische Hochschule*). I visited the areas where Pauli had lived, the restaurants and bars where he used to go, and the streets he used to walk, and stood outside his home in Zollikon, just outside Zürich. It was a large, nondescript, suburban detached house surrounded by trees, not the grand house I had imagined.

In Hamburg I walked the streets where Pauli had lived, worked, and played. Some of the bars he frequented in the Sankt Pauli red-light district are still there and still carry the same edge of violence.

At *La Salle Pauli* at the huge nuclear physics research laboratory

CERN (*Conseil Européenne pour la Recherche Nucléaire*), outside Geneva, where Pauli's library is housed, I looked through his books, marked in his own handwriting with his code for important passages, both books he read before meeting Jung and during the time he knew him.

Jung's Gothic mansion, two stops on the train from Zollikon, was, I had heard, no longer open to visitors. Nevertheless I sent a letter there, addressed simply to "The resident of 228 Seestrasse." A few days later I received an email from Jung's grandson Andreas, inviting me to visit. A gracious and friendly man—and the spitting image of his grandfather— he showed me around Jung's vast and splendid residence. I was thrilled to step inside the spacious high-ceilinged library where Jung and Pauli used to sit, first as patient and analyst and then as friends, mulling over the mind, the times in which they lived, and the civilization they knew. I looked around the dining room and put my hand on the table where they had dined. Outside the grand windows the lawn stretched down to Lake Zürich. It was the same view that the two friends used to admire as they chatted over fine wine and fine tobacco.

The table in Jung's dining room, the seventeenth-century alchemical books in his library, and Pauli's own books, with his markings, brought home to me the intensity of their common quest. For Pauli realized that quantum mechanics—despite its grandeur, and in the face of his distinguished colleagues—lacked the power to explain biological and mental processes, such as consciousness. It was not a complete theory. As he put it, "Though we now have natural sciences, we no longer have a total scientific picture of the world. Since the discovery of the quantum of action, physics has gradually been forced to relinquish its proud claim to be able to understand, in principle, the *whole* world." To Pauli the only hope was an amalgam of quantum mechanics and Jung's psychology.

Jung's and Pauli's was a truly unique meeting of the minds. It was, as Jung wrote, to lead both of them into "the no-man's land between Physics and the Psychology of the Unconscious . . . the most fascinating yet the darkest hunting ground of our times."

Deciphering the
Cosmic Number

1

Dangerously Famous

IN 1920s EUROPE, Carl Jung was a celebrity and regarded as the chief rival of the great Sigmund Freud. While Freud had carved out the new field of psychoanalysis, it was Jung who made it fashionable. He extended the boundaries by using dream images to explore the unconscious more deeply than Freud had, probing into the archetypes built into our minds. He was a spell-binding lecturer and recipient of adulation both from colleagues and a host of women whom he referred to as his "fur-coat ladies." The rich and famous flocked to his fortresslike mansion on the shore of Lake Zürich, not only as prospective patients but also to enjoy his inspiring conversation. Among them were the McCormicks of the Chicago newspaper dynasty, H. G. Wells, and Hugh Walpole, who remembered him as looking "like a large genial cricketer." Some came just to gaze at the "primitive" who washed his own jeans with his "powerful arms" on the lawn outside his mansion.

Jung was, as he said himself, "dangerously famous," so much so that patients sometimes had to wait a year for an appointment. Psychoanalysis had become all the rage and "going to Jung was somehow very chic and modern," as a wealthy American female client put it.

But there was still something missing. Jung was concerned that his approach to psychoanalysis needed a scientific underpinning, but he didn't have the requisite scientific background. To develop his ideas, he needed to work with someone who was au fait with the latest developments in science.

Boyhood

Jung was born on July 26, 1875, in the village of Kesswil on Lake Constance, on the northern border of Switzerland, to an impoverished Protestant pastor with a passion for learning. His mother, Emilie, had had three stillborn children before young Carl's arrival and had withdrawn into a world of ghosts and spirits. Jung's father moved from parish to parish but nothing seemed to help her. This often enraged him, leading to violent arguments between the two, during which young Carl would take refuge in his father's book-lined study.

Embarrassed by his shabby clothes and poverty, the boy for the most part kept away from others. His main interest was in his own rich dreams, in ghosts, in stories of the supernatural, and in séances. His charmed solitude came to an end with the birth of a sister, when he was nine. From then on he had to share with her the little attention he received from his parents.

Young Carl spent long periods of time staring at a stone and talking to it. One day he carved the top of a wooden ruler into a manikin and painted it to look like a village elder. He hid it in the attic, took it presents, and even wrote letters to it. Years later, he realized that what he had created was actually a totem—a primeval object of worship. It was a straightforward case of "archaic psychic components" entering "the individual psyche without any direct line of tradition." Thus he found in himself what he would later call the "collective unconscious."

By the age of eleven young Carl's brilliance was clear. He was also bigger and stronger than his classmates and always up for a fight. By fifteen he had read most of the books in his father's study, from adventure novels to Nietzsche's *Thus Spake Zarathustra* and Goethe's *Faust* (both "a tremendous experience for me," he recalled), as well as Kant, the Grail legends, and Shakespeare.

Jung's family was so poor that the only university he could go to was Basel, near enough that he could live at home. The question was what to study. He was interested in archaeology, but the university did not offer it. Then he had two dreams. In one he was digging up the bones of ancient animals, while the other concerned protozoa. From this he decided he should study some form of natural science. But if he studied zoology he would be bound to end up as a teacher. So he opted for medicine, even though his father had to petition for a stipend to support him. He started at the university in 1895.

Jung's calling

By his final year Jung had realized that his real interest lay in probing the secrets of the psyche: "Here, finally, was the place where nature would collide with spirit." Following up his childhood interest in dreams and ghosts, he wrote a dissertation entitled "On the Psychology and Pathology of the So-Called Occult Phenomena."

After a brilliant university career, Jung was immediately offered a position at the Burghölzli Mental Hospital by its director, the world-famous Dr. Eugen Bleuler, who had coined the term "schizophrenia." The hospital is in Zürich, sixty miles from Basel. Jung began work there in December 1900, when he was twenty-five. Physically so big that he dwarfed colleagues, handsome and brimming with enthusiasm, with a voice and laugh so loud they filled the room, Jung had a magnetic presence. He soon became the director's protégé.

A huge, sprawling, austere building, the Burghölzli loomed over Lake Zürich. To discourage thoughts of suicide it was built in such a way that none of the inmates could see the water. Inside, the building was sparse. Apart from the doctors' offices there were no comfortable chairs, only wooden benches. The working day at Burghölzli rarely ended before 10 p.m. Jung found the regime exhausting and missed the intellectual life of Basel with its late night philosophical conversations. But he was convinced that psychiatry was his metier.

Emma

Four years earlier, Jung had met Emma Rauschenbach, a fascinating fourteen-year-old girl from a very wealthy family. She was an heiress, the second richest in Switzerland. Her father owned a vast manufacturing empire that produced, among other things, machine-made watches, which were then a novelty. However, it was not Emma's wealth that attracted him to her. Jung always insisted most emphatically that he had not married her for her money. He once confided to a friend that he had fallen instantly in love the first time he saw her and felt sure they would marry some day. The friend, sadly to say, laughed.

In 1901 he was invited to a party in the town of Winterhur, where he met her again. She had grown into a beautiful young woman with dark hair, wide expressive eyes, and a ready smile. Having lived for a year in Paris, she spoke French and read Old French and Provençal. She was deeply interested in the legends of the Holy Grail, as was Jung. That summer her mother invited Jung to a ball at their elegant summer residence, the Ölberg, outside Schaffhausen. It covered several acres and there were scores of servants and gardeners. Jung's cardboard collar, tattered clothes, and rough manners were in sharp contrast to the well-heeled crowd. To add to his problems Emma was already betrothed to the son of a business associate of her father's.

Undaunted, he set about courting her. Emma was attracted to his good looks and brilliance. But more than that he seemed to value her intelligence and encouraged her to broaden it, qualities she found lacking in the other men who pursued her. In his numerous letters he suggested books she might read on subjects that included literature and psychology. She came to believe that more than just a wife to him, she could be a partner in his professional life. His courtship was boosted immensely by clandestine help from his future mother-in-law, who had come to realize her daughter's growing affection for the poor but affable Dr. Jung. Eventually she convinced her husband of the young man's seriousness and that her daughter's happiness was of paramount importance.

The two married in 1903 in a lavish ceremony held at the Swiss Reformed Church in Schaffhausen. Two days later, in the best hotel in

Carl and Emma Jung at the time of their marriage, February 14, 1903.

town, there was a sumptuous wedding banquet of twelve courses, each accompanied by the proper wine.

Emma took on the task of transcribing the voluminous notes Jung made during his hospital rounds. The following year they had their first child. Emma's substantial wealth gave Jung the freedom to pursue his own research and he quickly came to the attention of the international psychoanalytic community. By 1906 he had been appointed senior doctor at the Burghölzli, second only to Bleuler.

The fur-coat ladies

Jung had also started lecturing at the University of Zürich, where he held students spellbound. He ranged over not only psychiatry but also history, culture, mysticism, hysteria, and family dynamics, focusing particularly on women's problems. Many in his audience were wealthy women from Zürich's high society—his "fur-coat ladies." After lectures they would flock to talk to him.

Some of the fur-coat ladies began to invite the professor back to their homes for private consultations. There were no ethical guidelines in psychoanalysis in those days and the treatment sessions, often intense, sometimes ended in sex. Jung's flirtations began to get out of hand. But the more dangerous they became, the more they excited him. He bragged of his "heroic efforts" to keep his female patients at arm's length. The hospital community was small and gossip quickly spread.

Emma was painfully aware of her husband's infidelities. Eager to placate her, he resigned from the Burghölzli, allowed her to assume a larger role in his day-to-day work, and built a new house for their growing family in Küsnacht, outside Zürich. She was pregnant with their third child in another effort to bring the family closer.

Jung even analyzed Emma to convince her that the rumors of his infidelities were untrue. In fact this was far from the case. As he wrote to Freud, "the prerequisite for a good marriage is license to be unfaithful." Jung soon resumed his infidelities. Whenever Emma threatened divorce he would suddenly become incapacitated, claiming fatigue from overwork or suffering a serious accident such as falling down stairs. Emma would have to drop everything to nurse him back to health. Eventually she resigned herself to the situation.

Jung meets Freud

Jung first became famous for his word-association tests. In these he measured patients' responses to stimulus words such as "mother," "father," and "sad," and noted cases where they did not reply or hesitated before replying. He concluded that the slower the response, the more deeply the patient was delving into his unconscious. The speed and quality of the responses, he wrote, could be explained by the action of "feeling-toned complexes," which he later abbreviated to "complexes." These, he suggested, existed below the conscious level and could only be perceived when the patient's threshold of attention was lowered by using stimulus words.

Like Freud he claimed that experiments of this sort were proof of the unconscious. But while he agreed with Freud's hypothesis that there was a personal unconscious that was constructed out of worldly experiences, he was also beginning to sense the presence of a deeper shared unconscious that could only be imagined.

His conclusions seemed to square with Freud's concept of repression in which people kept their neuroses locked up in their unconscious. The role of the psychoanalyst, in Freud's view, was to root them out to improve the patient's conscious life. Jung disagreed with Freud's hypothesis that sexual trauma was the primary cause of repression. He was eager to meet Freud and talk through these points of difference and in 1907 arranged to visit him in Vienna.

On that first historic meeting, Jung arrived at Freud's apartment for lunch at 1 p.m. and left some thirteen hours later. The main point of contention was Jung's belief that parapsychology deserved to be classed as a scientific field. Born in 1856, Freud was nineteen years older than Jung. He was only five foot seven, but his immaculate grooming, sharply observant eyes, and ever-present cigar gave him an aura of authority.

Freud's eldest son Martin was present and recalled vividly Jung's "commanding presence. He was very tall and broad shouldered, holding himself more like a soldier than a man of science and medicine. His head was purely Teutonic with a strong chin, a small moustache and thin close-cropped hair." Jung dominated the meeting with stories of his cases and new ideas told with great enthusiasm. Freud was delighted. The man with whom he had been corresponding met his every expectation— except one.

Freud clung to the late-nineteenth-century view that the mind, like everything else, could be reduced to matter and understood through physics and chemistry, subjects that he had studied in some detail as a medical student, in a daring departure from the traditional curriculum. This approach—known as "positivism"—had been founded by the philosopher and physicist Ernst Mach. Mach claimed that science could only study data obtainable in the laboratory and which could ultimately be experienced with the senses, such as direct touch.

Freud had begun his career as a medical researcher in neurology. This work further supported his view of the brain as a mechanism. But it was the teachings of the then-master-neurologist Jean Martin Charcot in Paris on the use of hypnosis to study hysteria—itself seen as a disease of the nervous system—as well as the analytic methods of the Austrian physician Josef Breuer that launched Freud on the route to his ever-lasting fame. Having found that hypnotizing patients with hysteria was often unsuccessful, he tried a "talking cure," in which patients talked through their problems. Thus was psychoanalysis born. Most of the fundamentals had been worked out by Breuer, Charcot, and Pierre Janet, the famous French psychologist, among others. Freud's genius was in putting them all together and then devising ways systematically to study the unconscious.

Although Jung initially considered a career in brain anatomy, his experiences at the Burgölzli led him to return to his original interest in regions of the unconscious that Freud, with his positivistic turn of mind, dared not enter—the world of the collective unconscious with its archetypes. Parapsychology, he was convinced, was a way to probe it. Jung often turned the tables on Freud by arguing that Freud's views on sexuality were as unprovable as his own on parapsychology in any strictly scientific sense, especially the positivistic one.

Eight months after their first meeting he wrote to Freud, "I have been dabbling in spookery again." He could not resist any opportunity to remind Freud of his interest in parapsychology.

Mach himself once remarked acerbically on Freud and his school of psychoanalysis, "These people try to use the vagina as if it were a telescope so that they can see the world through it. But that is not its natural function—it is too narrow for that."*

The debate between Jung and Freud on parapsychology and sexuality continued for several years. Jung strongly believed that the human psyche was much too complex to be understood merely in terms of the libido. Conversely, when Jung visited Freud in the spring of 1909, Freud categorically rejected the entire field of parapsychology, which Jung so fervently believed in, as "nonsensical." He offered supporting arguments, Jung recalled, in "terms of so shallow a positivism that I had difficulty in checking the sharp retort on the tip of my tongue. While Freud was going on in this way I had a curious sensation. It was as if my diaphragm were made of iron and were becoming red-hot—a glowing vault."

---

*It was Wolfgang Pauli, Mach's godson, who passed the comment on to Jung, no doubt to his amusement. P/J [60P], March 31, 1953.

As the two were arguing, there was a loud report in the bookcase next to them. They both jumped back, fearing it would fall on top of them. "I said to Freud," continued Jung, "there, that is an example of a so-called catalytic exteriorization phenomenon"—a physical effect brought about by a mental thought. "Sheer bosh," replied Freud. He dismissed Jung's explanation of what had happened as occult nonsense. They never again discussed the incident.

That very evening Freud named Jung as his successor in the psychoanalytic movement and adopted him as an eldest son. A month later he wrote Jung accusing him of taking away all the pleasure of the moment and divesting "him of any paternal dignity."

Around this time Jung had been treating Joseph Medill McCormick of the McCormick newspaper dynasty, for alcoholism. When he suddenly managed to bring about a cure, he overnight developed a huge reputation in the United States. Later that year he was invited to lecture at a symposium at Clark University, in Worcester, Massachusetts. Freud too was invited to lecture and the two traveled together. As they sailed into New York harbor, Freud turned to Jung and said, "If they only knew what we are bringing to them." During the crossing to the United States the two were together for a week, and often analyzed each other's dreams. In the course of analyzing one of Freud's, Jung asked him for details of his private life. Freud replied, "But I cannot risk my authority!" "That sentence burned itself into my memory," Jung remembered. "Freud was placing personal authority above truth."

In the United States Freud had a miserable time and suffered anxiety attacks, but Jung fell in love with the New World. He traveled to Maine and New York, where he immersed himself in the Egyptian, Cypriot, and Cretan collections at the Metropolitan Museum and studied the tapestries at the Pierpont Morgan Library and the palaeontology collections at the American Museum of Natural History.

He also met up with the Harvard philosopher and psychologist William James, brother of the novelist Henry James. The two immediately bonded. They discussed parapsychology, spiritualism, faith healing, and other nonmedical applications of psychotherapy, all of which went well beyond Freud's view of the psyche.

Jung was convinced that Freud's focus on sex limited his intellectual

Clark University, 1909. Sigmund Freud is seated on the left; Carl Jung is seated on the right.

horizons. As far as Jung was concerned, Freud employed literal interpretations and so, for instance, "could not grasp the spiritual significance of incest as a symbol." For Jung sex played no more than a role in the psyche. He was more interested in its "spiritual aspect and its numinous meaning." Freud claimed that he was so far above sexual inclinations himself that he could judge a patient's sexual disorders objectively. Jung, however, had begun to suspect that Freud was involved in an ongoing affair with his sister-in-law and even claimed that she had confessed it to him.

The tempestuous relationship between the two masters of psychoanalysis seemed likely to end in disaster. Yet between them they brought psychoanalysis to the New World.

The collective unconscious

Both Freud and Jung worked by analyzing dreams. But while Freud did so to probe an unconscious that he postulated as being made up of every-day experiences, Jung was interested in dreams as the portal to something that transcended the individual—a shared or collective unconscious.

Jung's dreams were steeped in symbolic content. In one he was in a house, on the second floor. As he came downstairs he seemed to slip back in time. In the basement there were two skulls. Freud regularly inter-preted dreams in terms of two conflicting drives in our mental life—the life drive (which includes sex) and the death drive. He was convinced there was a secret death-wish in Jung's dream and suggested that it meant Jung wanted to murder someone. Driven into a corner, Jung jokingly suggested it must be his wife and sister-in-law. To his amazement Freud was greatly relieved for it appeared to support his analysis. For Jung, con-versely, the ground floor simply represented the first level of his uncon-scious, and so on into the depths. The dream of the house revived Jung's old interest in archaeology and symbolism. Crusaders and the Holy Grail began to enter his dreams.

Another dream that contained more than Jung initially realized was that of "the solar phallus man." A severely schizophrenic patient interned at Burghölzli claimed that he had had a vision of a phallus hanging from the sun. When it swayed it produced weather and also allowed God to spread His semen. In those politically incorrect days, all the doctors, including Jung, thought this hilarious. Then, one day, Jung happened to read a book on Mithraic mythology. It described a liturgy of exactly the sort imagined by the schizophrenic patient. Jung knew for certain that the patient had never seen this book, or any like it. It seemed to provide firm evidence of what Jung was to dub, in 1913, the collective unconscious.

Freud saw the unconscious as a storehouse of repressed emotions, thoughts, and memories, an arena where a day-to-day struggle took place among the id, ego, and superego with a strong sexual undercurrent. Jung, conversely, was interested in aspects of the psyche that could not be attributed to an individual's personal development but belonged to the deeper nonpersonal realms common to humankind. The collective unconscious is made up of archetypes, which Jung initially referred to as "complexes"—the "feeling-toned complexes" that affected the speed of a patient's response to his word-association tests. In this primordial state they belong to what he called the psychoid realm, that is, they are not part of the psychic realm, which is made up of the personal unconscious and the conscious. Archetypes are not inherited ideas but potentialities—latent possibilities. Their origin remains forever obscure because they exist in a mysterious shadow realm of which we will never have direct knowledge, namely, the collective unconscious.

Jung described archetypes as an invisible crystal lattice shaping thoughts in the same way that a real crystal lattice refracts light. An archetype can be charged with energy, or "constellated," by perceptions or thoughts from the personal unconscious and can thus be visualized through archetypal images or symbols. Thus the archetype can move from the psychoid realm to the conscious.

Archetypes are built into the mind. They are organizing principles enabling us to construct knowledge through an analysis of incoming perceptions. They influence our thoughts, feelings, and actions and determine whether a mind is sick or well.

Some years later Jung noted his "amazement that European and American men and women coming to me for psychological advice were producing in their dreams and fantasies symbols similar to, and often identical with, the symbols found in mystery religions of antiquity, in mythology, fairytales, and the apparently meaningless formulations of such esoteric cults as alchemy. Experience showed, moreover, that these symbols brought with them new energy and new life to the people to whom they came."

In 1912 Jung published *Symbols of Transformation*, in which he began to develop in detail the concept of the collective unconscious. It was the final break with Freud. Almost until then, Jung professed a great devotion to Freud, his teacher. But the debate over the role of sexuality never faded away with Jung seeking a psychoanalytic theory of a more general—a more transpersonal—sort. As for Freud and his circle, they turned on Jung, damning him as an occultist. They reviewed his book harshly and he lost many friends and colleagues. Jung was disappointed but at least he was now absolutely free to indulge in the "images of my own unconscious."

Remembering the totem of his boyhood, he now perceived it as "a little cloaked god of the ancient world, a Telesphoros such as stands on the monuments of Asklepios and reads to him from a scroll." The only place he could have seen such a cloaked god in reality would have been in his father's library, but there was certainly no such thing there. This was surely evidence that there were archaic components in the individual psyche.

A new psychology

What could these images mean? Where did they come from? Jung wanted to develop a psychological view that encompassed the metaphysical and the irrational. As to how to proceed, he was inspired by a passage from Goethe's *Faust* referring to "continuity of culture and cultural history." By looking into history he could find the elements that make up the mind.

He was also inspired by the writings of Schopenhauer and Nietzsche. A generation of young turn-of-the-century German-educated intellectuals were attracted to Schopenhauer's underlying philosophical message that optimism was naive. Only an outlook pervaded by pessimism could capture the essential nothingness of mankind, he wrote.

But Jung saw something more, particularly in Schopenhauer's view of the mind—that representations visible to us emerged from an underlying invisible world, the abode of the ultimate physical reality. Surely this meant that mental or psychic energy could be traced to nodes of energy in the unconscious, namely archetypes.

From 1913, Jung began to take a new approach with his patients. Instead of applying a theoretical framework to his analytic sessions, he asked patients to report their dreams spontaneously, gently prodding them to interpret the dreams and helping them understand their own dream-images. He himself began to dream fantasies populated by violent dwarfs, savages bent on killing him, Biblical figures, and Egypto-Hellenic characters fraught with Gnostic coloration. He walked and talked with his dream figures. What was going on, he asked himself. Was this science and, if not, what was it? It was a riddle he had to crack.

Eventually he concluded that the images in his dreams came from the collective unconscious and were transmitted to his conscious. In 1916 he transferred his extensive dream notes into what he called his *Red Book*, richly illustrated in the manner of a medieval manuscript. Jung permitted no one to see this book during his lifetime. When R.F.C. Hull, Jung's longtime friend and translator, read it after Jung's death he wrote, "Talk of Freud's self analysis—Jung was a walking asylum in himself, as well as its head physician."

That same year Jung painted his first mandala (a diagram, usually based on a circle or square, with four symbolic objects symmetrically placed—and a key archetype and ancient symbolic device in cultures around the world). It flowed from him, he wrote, but he did not under-

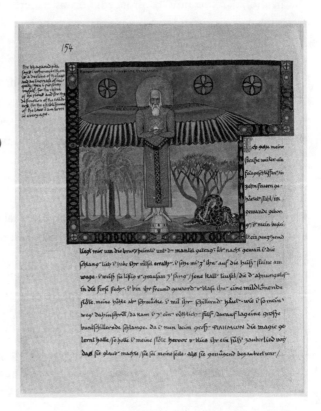

The old man whom Jung called Philemon and with whom he walked and had long conversations. (Jung, Red Book *[2009].)*

Jung's first mandala, drawn in 1916.

stand why he had painted it or what it meant. Art, it seemed, flowed from the unconscious. Convinced that his fantasies were spontaneous and self-created, he concluded that a mandala was a message that the conscious and unconscious had merged to become one: the Self was a whole. The appearance of a mandala in a dream signaled stability and inner peace, as Jung himself had come to feel.

The next step was to bring these insights together. He called his new version of psychology "analytical psychology" to distinguish it from Freud's psychoanalysis. But there was still a long way to go and many pieces of the puzzle to find.

2

Early Successes, Early Failures

THERE WAS something about Wolfgang Pauli. From early on in his career, colleagues couldn't help noticing that whenever he entered a laboratory, equipment spontaneously broke down. The Pauli effect, as it became known, was obviously impossible; it had to be just a matter of coincidence. But nevertheless it happened again and again.

There was the time, in the 1920s, that there was a massive equipment breakdown in a laboratory at the University of Göttingen, in Germany. Early one afternoon, without apparent cause, a complicated apparatus for the study of atoms collapsed. Pauli wasn't even in the country. He was in Switzerland. At last said his colleagues, relieved, here was clear proof it couldn't be the Pauli effect. The professor in charge of the laboratory wrote a humorous letter to Pauli to tell him about it. The letter was sent to Pauli's Zürich address. After a considerable delay, a letter arrived from Pauli with a Danish stamp on it. Pauli had been on his way to Copenhagen, he wrote. At the precise moment when the equipment broke down, his train had stopped for a few minutes at Göttingen station.

There were other things about Pauli too. He just happened to be one of the most brilliant physicists of his day. The great Einstein spoke of him

as his successor. But Pauli was extremely modest and refused to participate in the feverish competitive scrum in which all the other physicists
of his day pursued their research. Instead of publishing papers, he often
revealed his insights simply in letters and discussions. When others published papers laying claim to the same insights, he didn't bother to point
out that he had discovered them first.

He was also famous for the excesses of his private life. While other
physicists focused exclusively on their research, he'd spend his nights out
on the town, hanging out in bars and cafés, getting drunk, carousing
with singers and cabaret dancers, and getting into fights. He was never
up before midday.

He didn't look like a man who enjoyed the dark byways of life. In
fact he always seemed rather staid. He was short and plump with a round,
inscrutable face that colleagues said reminded them of a Buddha, and he
had a famously sardonic sense of humor. But everyone agreed he was a
wild character—though few people ever guessed that his life would fall so
out of control that he would end up going into analysis with Carl Jung.
Nor would anyone ever have imagined the effect this would have on his
life and thinking.

Boyhood

Wolfgang Ernst Friedrich Pauli was born on April 25, 1900, in Vienna
into a prestigious scientific family. His father, Wolfgang Josef Pauli, was a
chemist at the University of Vienna and had an international reputation.
Pauli's godfather, from whom he took his middle name, was Ernst Mach,
the famous physicist and philosopher.

On the occasion of Wolfgang's baptism, Mach presented him with
a goblet that he was to keep for the rest of his life. Many years later he
wrote of the powerful influence that Mach exerted on him:

> Among my books sits a somewhat dusty case containing an art nou-
> veau silver goblet in which lies a card. Now there appears to me to
> rise from this goblet a serene, benevolent, and cheerful spirit [Mach]
> from the bearded age. . . .

It so happened that my father, then intellectually completely under Mach's influence, was very friendly with his family. Mach had affably expressed his willingness to play the role of my godfather. He was, no doubt, a stronger personality than was the Catholic priest. The result seems to be that, in this way, I was baptized as "Antimetaphysical" instead of Roman Catholic. In any case, the card rests in the goblet and, despite my greater spiritual transformations in later years, I still label myself as being of "Antimetaphysical ancestry."

It is actually rather extraordinary that Pauli had a Catholic baptism, for his father came from a strongly Jewish family.

Pauli's grandfather, Jacob Pascheles, had been a leading member of the Jewish community in Prague, then part of the Austro-Hungarian Empire. His house was a meeting place for a religious group who called themselves the Paulas. In his synagogue, he presided over many bar mitzvahs, including Franz Kafka's. Pauli's father studied medicine at Charles University in Prague and joined the staff of the University of Vienna at twenty-three, in 1892. He went on to establish himself as a renowned expert in the chemistry of proteins. Six years later the government granted him permission to change the family name to Pauli and a year after that he converted to Catholicism. This was not unusual for a Jewish-born academic; in the anti-Semitic environment of the Austro-Hungarian Empire it increased his chances of academic advancement considerably.

Soon after converting he married Bertha Camilla Schütz, who was also Catholic, though in fact her mother's father was Jewish. She was highly intellectual. In the Vienna of the time, most women did not even attend high school. But Bertha was a correspondent for the influential liberal newspaper *Neue Freie Presse* for which she wrote theater reviews and historical essays.

Wolfgang was born a year after his parents' marriage. His family called him "Wolfi." His grandparents and aunts doted on him. He exhibited his love of precision from an early age. Once he was on a walk with his aunt, Erna. "See Wolfi," she said, "we are on a bridge crossing the Danube Canal." Four-year-old Wolfi set her straight: "No, Aunt Erna, this is the Vienna Canal which flows into the Danube Canal."

Wolfi was seven when his sister Hertha Ernestina (also named after

Mach) was born, in September 1906. (Later in life, perhaps to appear younger than she was, she was to list her birth date as 1908 or even 1909.) Pauli later confided to his second wife, Franca, that he had been jealous of the new baby and had felt rejected by his mother. But he seems to have got over it. In general theirs was a happy childhood. They grew up in a pleasant house on the outskirts of Vienna where they enjoyed exploring the surrounding woods and swimming in the Danube. Hertha recalled fondly how at Christmastime the family used to gather around the tree while Wolfi played "Silent Night" on the piano.

Young Wolfgang even tried to teach Hertha science. He showed her his collection of Jules Verne novels, precociously pointing out errors in the physics. When Wolfi was sixteen and Hertha was nine he decided to teach her astronomy. He informed his long-suffering sister that the so-called fixed stars were actually not fixed. In that case they must be falling, she replied. Pauli said no, but Hertha insisted. One can imagine the two arguing back and forth, the intellectual but blinkered young man and his equally stubborn sister.

It would have seemed there was not much point of contact between them, but in fact although they were never close, Hertha played an important role in Pauli's mental life, as he discussed with Jung. In later years the two maintained an affectionate relationship. Writing in 1959, Hertha reminded Wolfgang of the Aunt Erna story and signed the card, "I embrace you. Always, Your Hertha."

Pauli was ten when he entered the prestigious Döblinger Gymnasium in Vienna. He was popular with his fellow pupils, who remembered him as instigating many pranks. He had a talent for imitating professors. One, a particularly diminutive man, used to pop up unexpectedly amidst large groups of students. Pauli gave him the colorful and apt appelation *das U-boot* (the submarine). Four years later he had mastered geometry, calculus, and celestial mechanics, poring over books by the French polymath Henri Poincaré.

Among the odd events in Pauli's childhood, one of the oddest is his parent's conversion, when he was eleven, from Catholicism to Protestantism. Whether or not there was much discussion of the subject, no letters or records remain to explain this decision or the effect it might have had on the young Pauli.

Then, when he was sixteen, his father's mother made a rare visit. She revealed to him that his father's name was not Pauli but Pascheles and that he was Jewish.

This is Pauli's story. But the physicist Paul Ewald, who was the assistant to Arnold Sommerfeld—soon to become Pauli's mentor—has a different version. He recalls that when he first saw Pauli, he thought immediately that he looked Jewish and told him so. Pauli denied it. Ewald advised him to go and look in the mirror. When Pauli went home for a visit he asked his mother and father about their backgrounds and in this way found out about his Jewish ancestry. Of course any discussion of this issue was almost certainly taboo in the Pauli family. To further obfuscate his Jewish origins, Wolfgang Sr. often went to such extremes as to enter into polemical exchanges over a certain discovery with a chemist named Pascheles—who was in fact himself.

Pauli never spoke of the events surrounding his father's name change and his own discovery that he was Jewish. His widow, Franca, considered the change of name to be a family secret and became angry when interviewers mentioned it.

Nor did Hertha ever speak of it. In fact, taking the exact opposite stance to her brother, she insisted on speaking of herself as "half-Christian," and went on to write books for children about the lives of Catholic saints.

For some years Pauli was not altogether happy to have discovered his Jewish ancestry. He found himself exposed to virulent anti-Semitism in Germany and Austria, fanned to even greater heights by rumors that the Great War (World War I) had been lost because of Communist and Jewish conspiracies.

One year after his mother's death, in 1928, he decided to leave the Catholic church. Despite the fact that Judaism passes down the mother's line and his sister Hertha always insisted she was half-Catholic, not half-Jewish, Wolfgang decided he was Jewish. Could he have been seeking a community, although he never did join one? In any case, a little over a decade later to assert his Jewishness in Germany or Austria would have been suicide. He was only able to fully come to terms with his Jewishness after lengthy analysis with Jung. Shortly after that, Nazi racial laws made it impossible for him—and for Hertha too—to avoid it. "With me everything is complicated," Pauli later wrote.

Sadly, there is little personal correspondence on the subject—or on any other subject—among Pauli, his sister, his mother, and his father. There must have been letters and perhaps they are still out there. In the meantime, the closest we can get to the sort of man Pauli was is through information from those who knew him well.

At the gymnasium, Pauli quickly began to find his science classes too easy for him. By now he was receiving private tutoring in advanced physics from professors at the university and was soon discussing his ideas with them. He relieved his boredom by studying Einstein's latest papers on the general theory of relativity, which he kept hidden under his desk, and was soon in full command of the theory.

Einstein discovered the general theory of relativity in 1915. It explores the motions of objects, from rocks to planets, stars and galaxies, and asserts that the world in which we live actually has not three but four dimensions—the usual length, breadth, and height, fused with time. The surface that light rays course over is a four-dimensional one, sculpted by the objects on it, in the same way that objects resting on a rubber sheet create wells. The indentations on this surface cause objects to roll toward each other and are experienced as gravity. Einstein's general theory of relativity may have started out as one man's view of the universe, but it was turning out to be the correct one. Scientists still refer to it as the most beautiful theory ever formulated.

At the time most physicists could not fully understand the theory's elegant mathematics and wide-ranging concepts. Those who could realized that there was still a great deal of work to be done on it. This included clarifying and broadening aspects of Einstein's mathematics, applying the theory to special cases—no mean task bearing in mind the abstruse mathematics—and widening it to include electricity. Pauli, at the tender age of seventeen, dived right in.

By this time Pauli was reading mathematics and physics books until two in the morning, like novels. Usually physicists read and reread textbooks, marking them with notes and underlinings. Most of Pauli's books were unmarked. He only needed to read them once.

Pauli excelled in the courses that interested him—mathematics and physics—but had mediocre grades in nonscience subjects. Later in life he insisted that he had enjoyed them, especially Latin and Greek. In the annals of his school, Pauli's class is remembered as the "class of geniuses," and Pauli as the chief genius. Of the twenty-seven boys in Pauli's class, two went on to win Nobel Prizes, while others made their name as actors, orchestra conductors, university professors, and leaders of industry.

Pauli graduated in 1918 with distinction and a mere two months later submitted a paper on relativity theory for publication. In it he clarified certain fine points of a particular version of Einstein's theory, extended to include electricity (this had recently been proposed by the renowned mathematician and physicist Hermann Weyl in Zürich). In acknowledgment of his famous father, he signed this maiden paper Wolfgang Pauli Jr. It seemed the world was his oyster.

Meanwhile the Great War was coming to an end. Pauli was forced to undergo a medical examination to join the army and to his relief was found to have a weak heart, thus excusing him from conscription. He was passionately opposed to the war and the establishment, as classmates at the gymnasium recalled, and this only increased as the war went on.

Pauli was determined to pursue a career in physics, but as far as he could see Vienna was an intellectual desert, devoid of famous physicists. He decided to move to Munich, where the physicist Arnold Sommerfeld—a pioneer in the exciting field of quantum physics—was holding court.

According to one account, Pauli had already met Sommerfeld when he was a boy of twelve. The occasion was a lecture Sommerfeld was giving in Vienna and Pauli's father obtained permission for his son to attend. Afterward Sommerfeld sought out the youngster and asked whether he had understood the lecture. Pauli replied that he had, except for one equation on the upper left-hand side of the blackboard. Sommerfeld looked up and replied, startled, "There I have indeed made a mistake." True or not, the story attests to young Wolfgang's reputation.

Pauli in Munich

The young man who arrived in Munich in October 1918 was a sensitive-looking eighteen-year-old with hooded eyes, full lips, and a rather sullen look to his face. He wore his dark wavy hair combed back. He was not particularly tall, about five foot five, but he carried himself with an air of confidence.

Munich was a city in chaos. Germany, Austria-Hungary, and Turkey were on the verge of defeat and Allied armies—the Americans, English, and French—were poised to invade Bavaria (Munich was its capital city). The Allied blockade, exacerbated by two successive years of crop failures in the region, had resulted in starvation conditions. In 1917, in response to the huge losses of worker-soldiers, a radical minority of Socialists claimed that they had been double-crossed by imperialist capitalists. Inspired by the successful Russian revolution that same year, they agitated for a German Bolshevik revolution and the establishment of a Soviet-style Republic.

On November 7, 1918, a month after Pauli arrived, the Socialists organized a massive peace demonstration that was attended by over 50,000 people. With his strong socialist leanings, shaped by his mother's politics, Pauli undoubtedly sympathized and perhaps even joined in.

The Socialists demanded the abdication of the Bavarian King, Ludwig III, and proclaimed a Soviet Republic. Armed soldiers and civilians occupied key points in the city. Informed of all this, the king calmly packed his family into his new Mercedes and drove away. Despite its name, the Soviet Republic was a liberal regime and enjoyed widespread support.

Arriving in this chaotic city, Pauli found lodgings near the university at Theresienstrasse 66. Some of the buildings on this broad tree-lined avenue still stand and have been returned to their original condition. Like them, number 66 was probably in the classical style, built of earth colored bricks, and four floors with large windows curved at the top. Pauli's apartment was on the second floor, facing a courtyard at the back, which made the rent somewhat cheaper.

Turning left, Pauli would have strolled along Theresienstrasse and turned left again onto the Ludwigstrasse, a long straight boulevard, built in 1817 at the behest of King Ludwig I to reflect his love of Italy. Designed by an Italian architect, it resembles Rome's magnificent Via de Corso. Another few steps would have brought him to the university. Crossing the boulevard he would reach the main entrance with its piazza and massive fountain. Now, as then, students congregate there.

Arnold Sommerfeld, Pauli's new mentor, had studied at the University of Königsberg and immediately afterward had been called up for

Sommerfeld, about 1916, shortly before Pauli came to study with him.

military service. Unlike his fellow academics, he revelled in it and ever after sported a magnificent turned-up waxed mustache, which more than made up for his short, squat build.

Pauli had come to sit at his feet. But after a brief conversation, Sommerfeld concluded that he had little to teach the young man. "I have with me a really astonishing specimen of the intellectual elite of Vienna in the young Pauli . . . a first year student!" he wrote to a colleague in Austria.

Sommerfeld was impressed with Pauli's youthful paper on general relativity, which by now had been accepted for publication. It established Pauli as an expert in the field. He gave lectures on it at Sommerfeld's institute and the paper had even come to Einstein's attention. Pauli quickly wrote two more papers on relativity based on Weyl's research. Weyl now went so far as to speak of him as a colleague. "I find it impossible to understand how you, being so young, managed to acquire the means of knowing and freedom of thought [necessary to understand relativity theory in its entirety]," he wrote to the young man.

At the time Sommerfeld was writing an article on relativity theory for the *Encyclopaedia of the Mathematical Sciences*, of which he was editor.

Pauli attended a series of lectures he gave on the subject and offered some critical comments that intrigued Sommerfeld. So Sommerfeld asked him to co-author the article. In the end Sommerfeld became so heavily involved in his work on atomic physics that he turned the entire project over to Pauli.

The article was published in 1921 and was a tour de force. It was also published in book form and continues to this day to be an invaluable description of relativity theory. Einstein himself was impressed. "No one studying this mature, grandly conceived work would believe that the author is a man of twenty-one. One wonders what to admire most, the psychological understanding for the development of ideas, the sureness of mathematical deduction, the profound physical insight, the complete treatment of subject matters [or] the sureness of critical appraisal," he wrote in a review. Not long afterward, the precocious Pauli was addressing letters to giants such as the British scientist Arthur Eddington. Eddington was exploring a part of the mathematics of the general theory of relativity that needed some firming up: how to connect points on the curved surface of its four-dimensional geometry. Pauli informed him that his results were at "the moment meaningless for physics."

Around this time Einstein was to give an important lecture on relativity theory at the University of Berlin. Professors sat in the front row with lecturers and assistants behind them and the students at the back. Pauli, who was certainly not a professor and most probably still a student, chose to sit right in the middle of the front row dressed, according to one version of the story, in Tyrolean leather shorts. At the end of the lecture there was a hushed silence as the bearded professors in their starched white shirts and black ties decided who should ask the first question and what the order of precedence should be. Pauli sprang up, turned to face the audience, and announced with breathtaking chutzpah, "What Professor Einstein has just said is not really as stupid as it may have sounded."

Generally Pauli preferred to study on his own and attended classes only when necessary. One was the laboratory course that he was forced to attend with a friend whom he referred to as his "laboratory fellow sufferer." But he took an active role in the institute and was always on hand to discuss new developments in physics—in the afternoons, that is, for by now he had discovered the nightlife of Munich.

In the evenings, instead of turning left on Theresienstrasse to go to the university, Pauli would turn right. The road took him to the bohemian Schwabing district teeming with cafés, bars, beer gardens, and a huge number of cheap apartments. It was like a fusion of the Latin Quarter in Paris, where debates over the latest trends in art, music, and politics took place, with Montmartre, populated by the creators of these movements who generally lived in squalid conditions. Like the avant-garde scene in Paris there were publishing houses turning out new-wave literary magazines and satiric journals. It was the locus of radical experiments in art, politics, and sex. In Schwabing, anything went.

The area assimilated a spectrum of people. Before the war Vladimir Ilyich Lenin lived there while on the run from the tsar's secret police and schemed with associates such as Rosa Luxemburg. Among other denizens were the writers Thomas Mann and Rainer Maria Rilke and the artists Wassily Kandinsky and Paul Klee. It was a place where up-and-coming artists aspired to achieve their first success. Among these was a young man who arrived in 1913 and made a living by selling his paintings of the area to tourists. He tried to fit into the bohemian culture, but never really did. Twenty years later Adolf Hitler, in his autobiography *Mein Kampf* (My Battle), recalled the "Schwabing decadents."

Despite the upheavals in Munich, in October 1918 Schwabing was an island of calm. Café society was still bustling. Young Pauli, newly released from home, found it irresistibly attractive. He came to drink, to meet men and women, and perhaps to think about physics, while sitting at a table with a glass of wine or a cup of coffee. He had found the rhythm of his life.

Pauli spent longer and longer in Schwabing, though he was always careful to stay sober enough to be able to spend the small hours of the night working. It soon became impossible for him to make it to Sommerfeld's morning lecture, which began at 9.00 a.m. Instead Pauli took to dropping in at noon to check the blackboard to see what the topic had been so he could work it out for himself.

Sommerfeld reprimanded him. "In order for you to become a genius I have to educate you," he told him. "You have to come at eight o'clock in the morning." Unusually in the stiff world of the German Herr Professor Doktor, he was prepared to tolerate erratic behavior if the student was undoubtedly brilliant. Touched that Sommerfeld took such personal interest in his well-being, Pauli began to turn up at 8.00 a.m., at least for a while.

Pauli was particularly gripped by Sommerfeld's course on cutting-edge atomic physics, which focused on problems that the master himself was still struggling with. Fortunately for Pauli, the seminar took place once a week for two hours—in the evening.

After class a group of students often went to the café Annast, now part of the Hofgarten-Café, in the southernmost part of Ludwigstrasse, a short walk from the university's main entrance. For scientists the attraction of the Annast was its marble-topped tables, which provided excellent surfaces for scribbling equations during animated conversations. According to one story, Sommerfeld was once stuck on a particular equation. When he left the café he forgot to erase his attempts to solve it. The next evening he returned and found that another customer had solved it for him.

War zone in Munich

A short three months after Pauli arrived in Munich, this idyllic world of pondering the universe came to an abrupt end. Suddenly Munich was in the grip of anarchy. The moderate Soviet Republic formed just a few months earlier had lost the support of the populace. It was not surprising. Every day Pauli would have seen people standing hungry in the streets, lining up for food in the snows of one of the worst winters on record. A host of political factions sprang up and with great speed coalesced into two groups: a moderate to extreme right-wing group and a left-wing communist one. Both sides had no trouble recruiting an army. Central Europe was swarming with thousands of armed, disgruntled, and starving soldiers looking for a fight. The situation was ominous. News traveled fast around Munich that there had been a gun fight in the Bavarian diet, and that two representatives had been killed.

For sizable periods of time the university was shut down. Cafés became classrooms for Pauli and his fellow students and teachers. They also offered front-row seats for the street fighting carried on by uniformed soldiers as well as local citizens. Sometimes it was difficult to tell one side from another. By April there was a second Soviet Republic.

But this second Soviet regime also failed to contend with the food and fuel crisis and in April 1919 total chaos descended on Munich. Having suppressed an attempted communist coup in Berlin, the federal government dispatched an army to Munich to put down the last vestiges of rebellion against it. They blockaded the city, exacerbating the already critical food and fuel shortages. The struggle boiled down to a confrontation between the Communist army of the Soviet Republic—the reds—and the army from Berlin—the whites. The Red Army had 15,000 soldiers. The whites had about 40,000 soldiers, committed to eradicate by any means the Communists who they saw as a threat to the new republic.

Even walking the streets was risky because the reds were arresting and summarily shooting anyone suspected of spreading discontent or who looked suspicious.

The backbone of the white army—the Berliners—was the *Freikorps* (Free Corps). This was made up of extreme right-wing fascist paramilitary units manned by combat-hardened ex-soldiers serving as mercenaries, former officers often with royal titles and students who had been too young to fight in the war and sought instant action against easy targets. They were financed privately by German industry and hated Communists. One unit of the Free Corps called itself the defender of the democratic spirit against Communists and Jews. From it emerged such staunch "defenders of democracy" as Rudolf Hess, who was to become Hitler's deputy chancellor, and Ernst Röhm soon to command Hitler's storm troopers. It was also to provide the start in life for the man who was to become Pauli's closest colleague—Werner Heisenberg.

Preceded by a heavy artillery and mortar bombardment that created enormous damage and caused numerous civilian deaths, at the end of April the white army stormed Munich. Planes flew over, dropping leaflets telling people to surrender. There was heavy street fighting and massacres by both sides.

The fighting ended on May 8. Thousands of Red Army soldiers and civilian supporters had been killed—some estimates were as high as 20,000—in what became known as the white terror, wreaked mainly on the reds by the trigger-happy Free Corps. The white army estimated its losses at around 60. But the white terror was not yet over. People suspected of collaboration were summarily shot, stabbed, or beaten to death with rifle butts. Often they had been identified by spies who had infiltrated communist organizations, such as Corporal Hitler who had returned to decadent Schwabing. Munich, his favorite city, soon became the hot bed of his right-wing politics. The excesses of the Free Corps were so blatant that Lenin threatened to unleash Soviet forces on the area.

For Pauli, as for everyone, it must have been a traumatic and exciting time and also certainly dangerous. No doubt he wrote home about his experiences but sadly none of those letters remain. Perhaps Pauli's father destroyed them when he fled Vienna in 1938. Pauli, himself, may have destroyed others when he left Europe in 1940.

Pauli meets Heisenberg

By 1920 peace had returned to the city. Pauli was now Sommerfeld's deputy assistant. Among the students whose homework he had to correct was a young man called Werner Heisenberg.

Heisenberg was destined to become one of the great names in the history of physics. Even as a boy he was immensely competitive. He was not a natural athlete but trained with great determination and became an expert skier, runner, and Ping-Pong player. Like Pauli, he breezed through his classes at school and spent much of his time reading on his own, almost exclusively mathematics.

He had the look of "a simple farmboy with short, fair hair, clear blue eyes, and a charming expression," his friends recalled. Heisenberg first encountered atomic physics at the age of eighteen, in 1919, reading Plato's *Timaeus* while lying on a rooftop at the University of Munich during a break from his military duties as a member of the Free Corps, while rioting went on below him. (Five decades later Heisenberg was to recall those days as youthful fun, like "playing robbery [cops and robbers] and so on; it was nothing serious at all." Perhaps. Or perhaps not.) Heisenberg was entranced with Plato's description of atoms, visualized as geometrical solids. He knew this was now fantasy but was struck by the way in which the ancient Greek scientists were prepared to consider even the most unlikely speculations.

He had developed a keen interest in the theory of numbers and a year later entered the university. He had also tried studying Einstein's relativity theory. The reigning power in the mathematics department, Professor Ferdinand von Lindemann, convinced that Heisenberg's brush with relativity theory had spoiled his mind for a career in mathematics, rejected him outright. Sommerfeld, on the other hand, delighted with his enthusiasm and obvious brilliance, sent him straight to his graduate-level seminars, plunging him into advanced quantum physics.

Heisenberg and Pauli quickly struck up a friendship, cemented by their mutual passion for physics—although the two young men had diametrically opposite tastes when it came to what constituted a good time. Heisenberg recalled: "While I loved the daylight and spent as much of my free time as I could mountain-walking, swimming or cooking simple meals on the shore of one of the Bavarian lakes, Wolfgang was a typical night bird. He preferred the town, liked to spend his evenings in some old bar or café, and would then work on his physics through much of the night with great concentration and success." It was often said that in Germany just after the war, in the pre-Hitler years of the Weimar Republic, there were two types of people: those who went in for night life and those who dedicated themselves to the youth movement. Pauli typified the former, Heisenberg the latter.

Whenever they were apart, they corresponded, though their letters were more like scientific articles as they bounced ideas off one another. Just as he had corrected Heisenberg's homework in Münich, so Pauli continued to comment critically on Heisenberg's ideas. "Pauli had a very strong influence on me," Heisenberg recalled. "I mean Pauli was simply a strong personality. . . . He was extremely critical, I don't know how frequently he told me, 'You are a complete fool,' and so on. That helped me a lot."

"When I was young I believed I was the best formalist of my time," Pauli said later in life, referring to his extraordinary understanding of mathematics and how to use it in solving problems in physics. Mathematics had served him well in his papers on relativity theory. Now Pauli was to apply his mathematical acumen to another puzzle that he was determined to crack and which formed the subject matter of his PhD thesis. It related to the great Danish physicist Niels Bohr and his seminal model of the "atom as universe."

Niels Bohr and his theory of the atom

Bohr was another scientific prodigy. He arrived in England from Copen-hagen in 1911, when he was twenty-six, and became fascinated by the work of Ernest Rutherford at Manchester University. Rutherford had just unraveled the structure of the atom in a series of experiments that suggested that the atom was made up of a nucleus with a positive charge, surrounded by enough negatively charged electrons to produce an elec-trically neutral atom.

In other words, it was a sort of miniature solar system. But the model was unstable. Science was out of step with nature.

Bohr set out to solve this problem. He showed that electrons in an atom could not revolve in just any orbit—like planets—but that only certain orbits were allowed. Given that atoms are generally stable, their planetary electrons cannot be pulled into the nucleus. If they did then the atom would collapse. Bohr interpreted the stability of atoms as proof that there had to be a lowest orbit. He found it by altering Newton's theory of planetary motion using Planck's constant.*

In his atomic "bookkeeping" Bohr assigned to each allowed orbit a whole number, which he called the "principal quantum number." The lowest orbit was number one. As the principal quantum number

*Planck's constant—6.63×10^{-34} Joule-seconds (a Joule is a unit of energy, just as a mile is a unit of length)—is a measure of size of the discrete packets, or quanta, of energy emitted in the world of the atom. The significance of its smallness is indicative of the small scale at which quantum effects occur.

This measure lay at the core of the new quantum theory that Max Planck discov-ered at the turn of the twentieth century. The speed of light—three hundred million meters per second—was the signature of relativity theory, which describes nature in the large. At the other end of the scale, Bohr demonstrated how Planck's constant shaped the world of the atom—nature in the small. Planck's constant determines the Planck length—1.61×10^{-35} meters—the smallest conceivable length in physics. It is the length at which quantum effects take over. The Planck length is a combination of Planck's constant, the universal gravitational constant from Newton's law of gravity (6.67×10^{-11} (meter)^3/[kilogram $\times 39$ (second)^2]), and the velocity of light. To get some idea of its smallness, if a single atom were expanded until it was the size of the entire universe, the Planck length would be just four feet long.

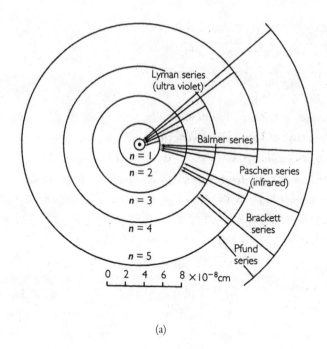

(a)

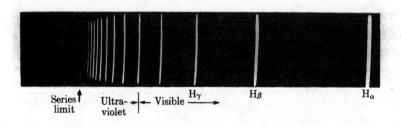

(b)

(a) This figure shows the hydrogen atom with its single orbital electron from Bohr's theory of the atom, where n *is the principal quantum number tagging the electron's permitted orbits. When the electron moves from a higher to a lower orbit there is a burst of radiation, and the frequency of this emitted radiation can be measured as a spectral line. The Lyman series, Balmer series, etc. are series of spectral lines. (b) This figure shows the Balmer series.*

increased, the orbits became closer and closer. Bohr called allowed orbits "stationary states."

Since the late nineteenth century scientists had been aware that when light illuminated a collection of atoms, they emitted light in response. When the light the atoms emitted was passed through an instrument that separated its frequencies—a spectroscope—lines appeared. Dubbed spectral lines, these lines were unequally spaced and bunched up more and more as their frequency increased. Most strikingly the series of lines were different for each sort of atom. In fact, an atom's spectral lines were its fingerprint, its DNA. Scientists had made a stab at writing equations to describe these lines, but there was no theory of the atom to explain the equations. Bohr's was the first to succeed.

According to Bohr's theory atoms emitted light when an electron moved from an upper to a lower orbit. The light that was emitted by an electron had the same frequency as a spectral line that had been observed. An oddity of the theory was that the electron's transition from one orbit to another could not be visualized—it disappeared and appeared again like the Cheshire cat's smile. In this sense the electron's quantum jumps were discontinuous.

Bohr's was a magnificent theory and it worked more than adequately. When applied to the hydrogen atom, the difference between the spectral lines observed in the laboratory and the spectral lines deduced from his theory was only 1 percent.

Scientists were impressed not only by its accuracy but also by its iconic visual imagery: the atom as a miniscule solar system with the electrons revolving in circular orbits around a central "sun," or nucleus. It was a momentous fusion of large and small, of the universe and the atom, the macro- and microcosmos.

Bohr's theory depicted the simplest element, hydrogen, as a single electron orbiting a positive charge—its nucleus. The atom had no total electrical charge; it was electrically neutral, just as atoms are in nature. Helium, the next element in the periodic table of the chemical elements, differs from hydrogen in that it has two electrons that orbit around a nucleus that has two units of positive electric charge. Because helium does not react chemically—it cannot bond with any other element—Bohr deduced that the innermost orbit needed to be filled up with two

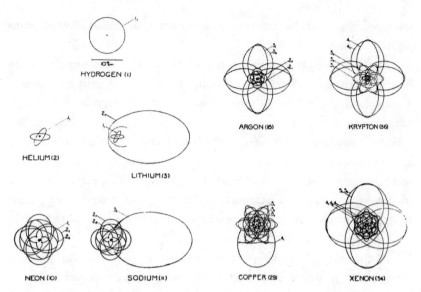

Atoms depicted according to Bohr's atomic theory. (Kramers and Holst [1923]).

electrons. He went on to infer that the next orbit can take on eight electrons.

As another of the pioneers of atomic physics, Max Born, head of the Institute for Atomic Physics at the University of Göttingen, put it:

> A remarkable and alluring result of Bohr's atomic theory is the demonstration that the atom is a small planetary system. . . . The thought that the laws of the macrocosmos in the small reflect the terrestrial world obviously exercises a great magic on mankind's mind; indeed its form is rooted in the superstition (which is as old as the history of thought) that the destiny of men could be read from the stars. The astrological mysticism has disappeared from science, but what remains is the endeavor toward the knowledge of the unity of the laws of the world.

Pauli's work on Bohr's theory

No one had attempted to apply the Bohr theory to anything more complex than the hydrogen atom. Pauli set out to do so.

He began the year after he arrived in Munich and decided to apply Bohr's theory to the next simplest atomic system to the hydrogen atom, that is, two protons orbited by a single electron—the hydrogen-molecule ion, H_2^+. The mathematics he had to grapple with was extremely complex. The problem gnawed at him. It took over his life. He ended up thinking about it night and day.

The last thing Pauli wanted was to disprove Bohr's iconic model. But after two years of working on the problem he had to conclude that he had proved beyond doubt that Bohr's theory could not produce the necessary orbits—or "stationary states"—for a stable H_2^+ ion. When he applied Bohr's theory, he discovered that a small disturbance to the electron orbiting the two protons would make it fly away from them. But that couldn't be right because stable H_2^+ ions had already been found in the laboratory. This could only mean that there was something fundamentally wrong with Bohr's model—not at all the result Pauli had hoped for.

Pauli was deeply discouraged. But Sommerfeld praised his mathematical skill and thoroughness. Heisenberg considered Pauli's result ominous. "In some way this was the first moment when really this confidence [in Bohr's theory] was shaken," he said.

Then Pauli received an invitation from Max Born. Born had done important work in electromagnetic theory, relativity, acoustics, crystallography, and most recently atomic physics. He was highly impressed with Pauli's mathematical skills and invited him to spend six months at the institute. Pauli accepted.

"W. Pauli is now my assistant; he is amazingly intelligent and very able. At the same time he is very human and, for a 21-year-old, normal, gay, and childlike," Born wrote to Einstein. By Pauli's own account, he was actually rather miserable. Born and Pauli applied Pauli's mathematical methods to the helium atom (He)—two electrons orbiting a nucleus. But Bohr's theory failed here, too. It horrified Pauli that all his work seemed to result only in undermining this iconic theory. As far as he was concerned, it was he who had failed, not the theory. This failure loomed over him and grew into a general sense of gloom.

Despite Born's presumption of his "gaiety," he had also noted that Pauli "cannot bear life in a small city." Nor was Pauli particularly enamored of working with Born. While Born was neat and well organized,

Pauli was not. Born was an early riser, Pauli far from that, especially after late nights working. Born often had to send someone to Pauli's apartment at 10:30 in the morning to awake him for his 11 o'clock lecture. Born recalled: "Although a place like Göttingen is accustomed to all kinds of strange people, Pauli's neighbors were worried to watch him sitting at his desk, rocking slowly like a praying Buddha, until the small hours of the morning."

Pauli also did not appreciate Born's overly heavy mathematical style of physics. He felt that the time was not yet ripe for such a rigorous approach. For him adroit guesswork backed up by mathematics was the best way to proceed.

Three months after his arrival in Göttingen, Pauli was offered a position as assistant to Wilhelm Lenz, professor of physics at the newly established University of Hamburg. He immediately accepted.

When Pauli had first arrived in Göttingen, Born lamented to Einstein that, "Young Pauli is very stimulating—I shall never get another assistant as good." In fact he did. Two years later Heisenberg appeared. Born wrote to Einstein, "He is easily as gifted as Pauli, but has a more pleasing personality. He also plays the piano very well." The two tried again to tackle the helium atom using other mathematical approaches to solar system models and failed. "All existing He[lium]-models are false, as is the entire atomic physics," was Heisenberg's bleak assessment.

Pauli meets Bohr

In June 1922 Pauli returned to Göttingen to attend a *Bohr-Festspiele* (Bohr festival) Born had organized to launch his new Institute for Atomic Physics. Sommerfeld, too, was there along with Heisenberg.

"A new phase of my scientific life began when I met Niels Bohr personally for the first time," Pauli later recalled. For both Pauli and Heisenberg it was hugely inspiring to meet this giant of physics.

Fifty years later Heisenberg still remembered the mesmerizing way in which Bohr presented his theory. He "chose his words much more carefully than Sommerfeld usually did. And each one of his carefully formulated sentences revealed a long chain of underlying thoughts, of philosophical reflections, hinted at but never fully expressed. I found this approach highly exciting. . . . We could clearly sense that he had reached his results not so much by calculation and demonstration as by intuition and inspiration, and that he found it difficult to justify his findings before Göttingen's famous school of mathematics." As physicists used to say, at Munich and Göttingen you learned to calculate, but at Bohr's center at Copenhagen you learned how to think. This was certainly the case for Heisenberg and Pauli.

One of the ways in which Bohr's original theory had been developed was to allow electrons to move in three dimensions, transforming the orbits into shells. In his lecture he described his latest work, which concerned how the electrons in an atom distributed themselves. The essence of the problem was the way in which atoms were built up, beginning with the hydrogen atom with its single electron. Understanding this would be a first step to explaining why the periodic table of chemical elements fell into place as it did, a key problem that everyone recognized needed to be cracked.

Examining the experimental data, Bohr proposed that there were two electrons in the first shell, eight in the second, and so on. He was then able to deduce these numbers from a series, each of which could be obtained from the formula $2n^2$, where n is the principal quantum number for the shell (in Bohr's original model, n was the principal quantum number for an orbit, but orbits had now been replaced by shells). If n is 1 then the total number of electrons in the first shell is 2; if n is 2, the number of electrons in the second shell is 8; the next shell becomes 18, and so on. The assembled scientists discussed this series of numbers with great intensity. But as they listened, both Heisenberg and Pauli realized that there was no basis in fact to what Bohr referred to as the "building up principle." As for Sommerfeld, he dismissed Bohr's reasoning as "somewhat Kabbalistic."

Bohr's reasoning also failed to answer the problem of why every electron did not simply drop into the lowest shell—at least for atoms other than hydrogen and helium.

Pauli and Heisenberg argued fiercely with Bohr, who was impressed with their knowledge of physics and their uninhibited critical give-and-take. Bohr was also impressed with Pauli's work on the H_2^+ ion and the helium atom, although in each case he had come up with a result that seemed to disprove Bohr's own theory. Discussions continued into the evening in coffeehouses and on walks. Heisenberg complained that no one went to bed before 1 a.m. Pauli, of course, was in his element.

After the meeting, Bohr invited Pauli to visit Copenhagen for a year, from September 1922, to help in his research. Pauli replied rather arrogantly, "I hardly think that the scientific demands which you will make on me will cause me any difficulty, but the learning of a foreign language like Danish far exceeds my abilities." He accepted nonetheless.

In Copenhagen Bohr set a problem that was to haunt Pauli for years and was one of the factors that led to his breakdown. The problem was on the anomalous Zeeman effect. To understand it, we have to go back to Pauli's old mentor at Munich, Arnold Sommerfeld, and his work on the structure of spectral lines.

The discovery of 137

According to Bohr's theory, when an electron drops from a higher to a lower orbit it emits light, which is recorded in the laboratory as a spectral line. Bohr was able to work out equations for these spectral lines that could be compared with the data obtained in the laboratory.

By 1915 scientists had more accurate spectroscopes that enabled them to make closer inspection of spectral lines. This revealed that many of the individual lines were in turn made up of several more closely spaced lines: they were said to have a "fine structure." Certain spectral lines also split into several lines or "multiplets" when the atom was placed near a magnet, but the fine structure was always there. "It was given by Nature herself without our agency," Sommerfeld wrote.

Sommerfeld's primary contribution to atomic physics was his work on the fine structure problem. His brainwave was to apply relativity theory to Bohr's theory, changing the mass of the electron following Einstein's famous equation $E = mc^2$. The result was astounding: an extra term appeared in Bohr's equation for a single spectral line. This extra term made it possible to predict that certain lines would actually split and reveal their fine structure.

Sommerfeld called the quantity that set the distance between the split spectral lines in this extra term the "fine structure constant" and designated it with the Greek letter α (alpha). His equation was:

$$\alpha = \frac{2\pi e^2}{hc}$$

The fine structure constant is made of three fundamental constants: the charge of the electron e (1.60 × 10^{-19} coulombs—a coulomb is the unit of electric charge); the speed of light c (3 × 10^8 meters/second), which defines relativity theory; Planck's constant h (6.63 × 10^{-34} Joule-seconds), which defines quantum theory and determines the size of the grains into which the microscopic world is partitioned, be it grains of energy, mass, or even of space itself. π (pi) is the ratio of the circumference of a circle to its diameter (3.141529). The constants e, c, and h had already been measured. Thus the discovery of the fine structure constant was a step toward the great goal of finding a theory that would unite the domains of relativity and quantum theory, the large and the small, the macrocosm and the microcosm.

There was one extraordinary feature of the fine structure constant. The three fundamental constants that make it up have dimensions—such as space and time—and therefore depend on the units in which they are measured, whether metric, imperial, or some other. So although they would certainly play an essential part in a relativity or quantum theory formulated by physicists on a planet in another galaxy, they might not have precisely the same values as they have on earth.

But when they come together to form the fine structure constant, something extraordinary happens. All of their units cancel out and as a result the fine structure constant is a pure number without any dimensions. No matter what the number system this will always be true. Sommerfeld calculated it as 0.00729—a rather unexciting way of expressing such a momentous result.

Sommerfeld's extension of relativity into atomic physics was "a revelation," wrote Einstein. Bohr wrote to Sommerfeld, "I do not believe ever to have read anything with more joy than your beautiful work."

A dimensionless number of such fundamental importance had never before appeared in physics. Of course dimensionless numbers had always been present in equations, but never one that was deduced from fundamental constants of nature. Scientists later realized that if the numerical value of the fine structure constant were to differ by a mere 4 percent, almost all carbon and oxygen would be destroyed in every star in the universe and life on our planet would not exist or would be dramatically different. The fine structure constant was one of the primal numbers that bound all existence together.

One of the many puzzles that arose was the question of why spectral lines of atoms split when they were placed in a magnetic field, between the pole faces of a magnet. Back in the mid-nineteenth century the British scientist Michael Faraday had identified the phenomenon but his equipment was not yet precise enough to enable him to pursue it.

In 1896 Pieter Zeeman, a young Dutch researcher at the University of Leiden, was looking for a research problem. Going through physics journals from decades earlier, Zeeman came upon Faraday's ruminations over the behavior of atoms in magnetic fields. With the more precise equipment at his disposal, he succeeded in discovering the additional split spectral lines caused by a magnetic field. This was dubbed the Zeeman effect.

Two years later all was not well again. When Zeeman tried using a weaker magnetic field, he found that the spectral lines split into different patterns of even more lines—multiplets. This peculiar situation became known as the "anomalous Zeeman effect."

The puzzle that Bohr set to Pauli was to find an equation that described this behavior. The effect existed; it had been identified. Therefore it must be possible to deduce an equation from Bohr's iconic theory of the way atoms worked—electrons revolving like planets in small solar systems. Despite its shortcomings Bohr's theory offered the only means to deal with problems of atomic physics. Perhaps it could be modified to suit the one at hand.

Day and night Pauli thought about it. He calculated and calculated, he tried this approach and that approach, and eventually he fell into a fit of despair. Everything had been going so well. The boy genius's triumphant entry into Munich had been heralded by an important paper on relativity theory and two more quickly followed. Even Einstein had been impressed. But ever since it had been nothing but one failure after another: "A colleague who met me strolling rather aimlessly in the beautiful streets of Copenhagen said to me in a friendly manner, 'You look very unhappy,' whereupon I answered angrily, 'How can one look happy when he is thinking about the anomalous Zeeman effect?' "

There had to be a way. But how?

3

The Philosopher's Stone

Jung's analytical psychology: The four function types

MEANWHILE, in Zürich, Carl Jung was establishing a vocabulary and framework for his budding new field of analytical psychology. In 1921 he published his seminal book on the subject, *Psychological Types*.

In this he argued, based on his vast experience with patients, that there were two opposing modes of being that determined and limited a person's reaction to the world and to himself—introversion and extraversion. In Jung's initial definition, introversion is a turning inward from an object, while extraversion is the reverse. Jung was the first to coin these two terms, which have since become common currency. He then broke these two categories down further and proposed four basic functions or function-types: thinking, feeling, sensation, and intuition. He was intrigued that there appeared to be four function types rather than three or some other number. But for the moment he set the problem aside.

Jung defined what he called the four orienting functions of consciousness thus: *thinking* leads to logical conclusions; *feeling* is a means to establish a subjective criterion of acceptance or rejection; *sensation* directs

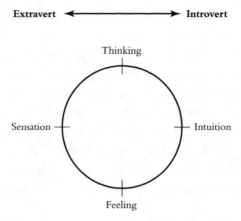

The two opposing modes of being and four function types.

one's attention outside oneself and is caused by conscious perception through the sense organs. As for *intuition*, it is somewhat like sensation but there is no cause for directing one's attention. Rather, there is a hunch, an inspiration, or gut feeling. Conclusions surface not by logical means but as if bursting out of nowhere, such as in suddenly realizing how to solve a problem when you are not consciously thinking about it.

Jung then divided these four functions into two groups of two: thinking and feeling, which are to do with rationality and logic; and intuition and sensation, which he classified as irrational, outside of reason. Besides his clinical experience, Jung drew upon his knowledge of Eastern and Western religions and of myths, philosophy, and literature to support his theory of types. In particular, he drew on the notion of pairs of opposites such as evil/good, darkness/light, matter/spirit, which he saw as emerging from deepest history—before Christianity, the Hebrews, the Egyptians, and the Chinese—and providing the energy for creativity and for life itself.

The extent to which these four functions predominate in an individual, Jung argued, gives each person a mode of being. Specifically, thinking types direct their mental energy toward thought at the expense of feeling, which disturbs the flow of logic; feeling types are governed by their feelings. Similarly, to understand a situation with one's senses—by sensation—requires concentration and focus, whereas trying to intuit a situation requires taking in its totality and flitting around it. These are opposites because no one can do both at once.

When one function is particularly dominant, the opposite one may lapse into the unconscious and return to its earlier archaic state. The energy generated by this inferior function drains into the conscious and produces fantasies, sometimes creating neuroses. One goal of Jungian psychology is to retrieve and develop these inferior functions. Jung was careful to point out that no one is strictly a thinking or feeling type. We are all combinations of the two types and the four functions. Our personality, or psychology, results from a struggle among these opposites for equilibrium.

At this time Jung had also begun to study the Gnostic writers, spurred on by his interest in myths. He was aware that Freud had derived his influential myth of the primal father and its effect on the superego from the Gnostic motifs of sexuality and Yahweh, which dated back to ancient Egypt and early Judaism. The Gnostics speculated that the content and images of the primal world of the unconscious might be clues to uncovering the mysteries of the universe. But at first Jung could find little relevance in their writings, nor could he find any historical bridge between Gnosticism and the contemporary world.

And still he dreamed. What could these images mean? Where did they come from?

Among the most vivid of his dreams were two in which he found himself in a huge manor house. In one he wanders from room to room and eventually ends up in a spacious library full of books from the sixteenth and seventeenth centuries. The engravings on the books are unfamiliar and the illustrations include curious symbols. In the second he is in a horse-drawn coach that enters a courtyard. Then the gates slam shut and a coachman screams that they are trapped in the seventeenth century. His efforts to explain this dream sent him delving into books on history, religion, and philosophy, particularly of that period.

Meanwhile, he was shaping his own method of treating neuroses. Freud interpreted a boy's incestuous desires for his mother as a literal return to the womb, to a state free from responsibility and decisions. Jung preferred to see the positive side, as breaking down the bond between mother and son and thus freeing psychic energy to be transferred to other archetypal components. In this way he removed the purely sexual connotation of incest, choosing rather to explore it in terms of archetypal metaphors and symbols. By now Zürich had become the center for this developing technique of analysis, Jung's "analytical psychology."

Alchemy

Back in 1914, Jung had come across a book by Viennese psychologist Herbert Silberer, who was part of Freud's circle. In *Problems of Mysticism and Its Symbolism*, Silberer discussed whether there might be a relationship between the imagery in alchemical texts, the imagery experienced by patients in the mental state between dreaming and waking, and Freud's analysis of dreams. At first Jung was fascinated and corresponded with him. He was looking for something deeper than Silberer—to understand the imagery that had never been conscious, the imagery in the deep or collective unconscious. But he soon concluded that alchemy was "off the beaten track and rather silly."

Nevertheless Jung began collecting ancient alchemical texts. Then, in 1928, his friend Richard Wilhelm sent him a copy of his translation of the thousand-year-old Taoist-alchemical text *The Secret of the Golden Flower*.

At first *The Secret of the Golden Flower* did not seem to make any sense. But Jung was intrigued. Silberer's book came to mind and he suddenly realized that although he had appreciated what Silberer was suggesting, he had not understood how to interpret the alchemical texts Silberer used. For the next two years Jung pored over alchemical texts and began to find more and more passages that he could understand. Then he had a revelation. "I realized that alchemists were talking in symbols—those old acquaintances of mine," he wrote. It was the symbols not the text that were the essence. He decided to learn alchemy from the ground up and then return to Silberer's and Wilhelm's books.

Alchemy was conceived of as a means toward understanding the "great chain of being"—in other words, all life—stretching from our "corruptible world" to heaven. There were two sorts of alchemist. Scientific alchemists, the forerunners of modern chemists and metallurgists, searched for ways to transmute base metals into gold and jealously guarded their recipes. The mystical school of alchemy, however, interpreted transmutation as a spiritual path to redemption. They considered their laboratory experiments to be part of an inner process of maturing while nurturing a contemplative attitude. Alchemy embraced the teachings of the Greek philosopher Proclus as well as mystery religions such as Zoroastrianism and the ancient cults of Isis, Mitre, Cybel, and Sol Invictus.

Alchemists postulated that everything, even metals, was made up of the four elements—earth, water, air, and fire—and that these four elements could be transformed one into another. They called this process of transformation the "circle" or the "rotation of the elements." The goal of alchemy was to bring about a union of all four elements to produce the mystical fifth element—the quintessence, or the legendary "philosopher's stone," the ultimate state of enlightenment. In alchemical books the four elements were represented by the four sides of a square. The philosopher's stone—referred to as the one, the perfection, and imbued with the power both to transmute base metals to gold and to transform man into the illumined philosopher—is represented by a circle. It is the light hidden in dark matter; it combines creative divine wisdom and creative power. Christians sometimes identified it with Christ, while Buddhists symbolized it as the jewel in the lotus.

The first step in creating the philosopher's stone was to obtain the *prima materia*, the basic material from which all metals are derived, "philosophical mercury"—Mercurius, known also by his Greek name, Hermes, symbolizing the universal agent of transformation as opposed to the vulgar physical mercury of the scientific alchemists. Mercurius is present throughout the process of transformation from its dark beginnings (as prima materia) to its triumphant end (as the philosopher's stone). In this way Mercurius participates in both the dark and light worlds.

Prima materia, in its turn, comes out of the union of male—sulphur (the hot, dry, active principle)—and female—argent-vive, or mercury (the cold, moist, receptive principle). In alchemical philosophy this union symbolized the wedding of man and woman, the *coniunctio* of King Sol and Queen Luna (sulphur and argent-vive). Sol (the Sun) is the male force of the universe, creative will. Luna (the Moon) represents the receptive female force, wisdom. The material world is generated out of sulphur (fire and air) and argent-vive (earth and water), that is, out of the four elements. Thus the conjunction of all these gives rise to the world of mysticism.

Alchemy and psychology

As he read more and more deeply in alchemical works, Jung realized that he had discovered the "historical counterpart of [his] psychology of the unconscious." Alchemy provided an unbroken historical link between the ancient Gnostics of first-century B.C. and the contemporary world. Its roots went back through Gnostic writings to Plato, Pythagoras, texts attributed to the magus Hermes Trismegistus of ancient Egypt (referred to as Moses in the Kabbalah), and ancient creation myths such as the Enuma Elish from seventh-century-B.C. Babylonia.

The Hermetic view was that after the fall humankind had divided into two states, the male and the female. The alchemical wedding returns man to the original Adamic state—to Adam—thus reconciling opposing forces and creating the highest wisdom, which is the philosopher's stone. The alchemical wedding releases the world-soul—the soul of the whole world—which had lain dormant until this reconciliation and which unites the souls of individuals and also of the planets, which are living entities and not merely matter.

Thus Jung finally understood the meaning of his two dreams about being trapped in the seventeenth century, the period when alchemy was at its height. Hereafter primordial dream images, which he saw as visual symbols of archetypes, began to play a central role in Jung's analytical method, along with ancient myths and religion.

Jung's associates warned him that he might be considered a charlatan if he dabbled in alchemy. If a scholar of Wilhelm's standing could publish a book on alchemy, Jung replied, then so could he. Furthermore, he was convinced that alchemical imagery and notions of transformation could provide another approach to understanding the psyche.

So Jung set to work to incorporate alchemy into his analytical psychology. One of his patients, Aniela Jaffé, later to become his personal secretary and collaborator, recalled a particularly startling yet productive analytic session. She was describing her problems with her mother when

Jung abruptly cut her off with the words, "Don't waste your time." He went to his bookcase and took down the *Mutus Liber*, an alchemical book from the seventeenth century that contained only images, no text, and they spent the rest of the session discussing the images. Looking back on this and similar sessions in later years, she recorded that they had a more lasting influence on her than any of those spent in conventional therapy.

Thus by incorporating alchemy into his analytic psychology, Jung began to evolve a dramatic new way to understand the unconscious.

4

Dr. Jekyll and Mr. Hyde

The pesky anomalous Zeeman effect

IN THE AUTUMN of 1923 Pauli left Copenhagen to go back to Hamburg. He had still made no progress with the anomalous Zeeman effect. The problem, to recap, was to find the correct equation to describe the spectral lines of an atom placed in a weak magnetic field. He worried at it like a dog with a bone. But no matter how hard he tried, he just couldn't crack it.

Pauli was growing more and more despondent. He had become friendly with Bohr's assistant, Hendrik Kramers, who promised to visit him in Hamburg to cheer him up. Then Bohr decided that Kramers must go to England with him. Pauli wrote telling Bohr how deeply offended he was by this decision and how much he had looked forward to seeing his friend, whose presence "would mean a great deal for me psychically." "I feel myself so unwell," he added. He had just returned and already he was writing to Bohr about how unhappy he was. His letter was in effect a cry for help.

Not long after he had settled back in Hamburg he gave his inaugural lecture. His subject was the periodic table of chemical elements, but his heart was not in it. He was all too aware that the most basic problem in understanding it had yet to be resolved: what was the reason that the shells of electrons in each atom filled in the way they did? He had a hunch that it was related in some way to the multiplets in the anomalous Zeeman effect. Surely it was all tied together. After all, the way in which the shells filled up with electrons determined the numbers of spectral lines.

In struggling to find a mathematical description for the effect, he began by reworking equations from the *normal* Zeeman effect—where Bohr's theory produced equations that agreed reasonably closely with the spectral lines that had been observed. Pauli's goal was to apply the new equations to the anomalous Zeeman effect. Sommerfeld had already made some progress along these lines.

To study the anomalous Zeeman effect, physicists focused on alkali atoms—primarily sodium, magnesium, and calcium—which displayed behavior similar to the hydrogen atom for which Bohr's theory seemed to work. Like the hydrogen atom, alkali atoms have only one electron in their outer shell and this is the only electron that can bond with other chemical elements. The other electrons are in the inner shells, which have already been filled and thus cannot react.

Sommerfeld set Heisenberg, who was just nineteen, to work on the problem. Since the Bohr theory now dealt with an electron in an atom moving on a three-dimensional shell, in addition to a principal quantum number each electron had two more associated with it. Locating an object in a room requires three numbers, two to give its location from the walls and the third its height from the floor. An electron can be located in an atom in a similar way, with three numbers identifying its position within the Bohr atom relative to the nucleus. These are taken to be whole numbers and are called quantum numbers.

Sommerfeld supplied Heisenberg with the newest data as well as his own unpublished research, including speculation on new ways to combine the three quantum numbers at the very basis of Bohr's theory of the atom. "All right, you have an interest in mathematics; it may be that you know something; it may be that you know nothing. . . . We will see what you can do," he said. Heisenberg quickly came up with his own ideas on how to tackle the anomalous Zeeman effect. He rewrote one of Sommerfeld's equations using half figures—1/2, 3/2, and so on—and discovered he could produce an equation that described most of the observed multiplets.

Then he turned to Bohr's model of the atom—with the nucleus as a rigid core surrounded by filled shells of electrons, the whole thing spinning like a ball. Heisenberg made the audacious assumption that the core and the surrounding electron shared a half unit of angular momentum by means of an interaction that he left unspecified. (An object moving in a line has linear momentum [mass times velocity]. Similarly an object spinning like a top has angular momentum [mass times angular velocity].) The mysterious interaction between the core and the lone electron could be the explanation for the anomalous Zeeman effect. Sommerfeld was stunned, as was Pauli. Surely this would result in an atom emitting a half quanta of energy. But that had to be wrong because quanta were assumed to be indivisible. This was a basic postulate of the quantum theory. All the same, Heisenberg's equation produced multiplets for the alkalis which precisely duplicated data from experiments. "Success sanctifies the means," Heisenberg wrote to Pauli.

Sommerfeld was astonished that this novice dared take such a dramatically different approach to problems with which experienced scientists had struggled. Instead of getting tied up in endless complicated calculations, Heisenberg came up with instant solutions. Eventually Sommerfeld had to give in and accepted that there had to be half quantum numbers. After all, he reasoned, classical physics was frequently proved wrong. Why not atomic physics, too?

Bohr, however, insisted that while breakdowns in classical physics were fine, it was not acceptable when it came to his own theory of the atom. At the Bohr-Festspiele in Göttingen, he had discussed Heisenberg's new approach and referred to it as "very interesting," by which he meant that it was almost certainly wrong. Although it happened to fit existing data, Bohr argued, it was not an end in itself. Bohr was more interested in unraveling a problem than in instant solutions.

Bohr now suggested that there might be a force that linked the core and the lone outer electron in an alkali atom and that this force might distort the core in two different ways, giving rise to a "double-valuedness," which he, too, was willing to include as a half quantum number. Thus

Bohr was able to reproduce the required multiplets, while avoiding the other half quantum numbers that were essential to Heisenberg's model. But what was this strange force? Pauli couldn't accept it and argued tooth and nail with Bohr. He continued to torture himself over the problem of the anomalous Zeeman effect but could make no sense of it. Bohr insisted that Pauli publish his own contribution to these mathematical models and he did so "with a tear in my eye," as he wrote to Sommerfeld. As for Heisenberg's theory of the anomalous Zeeman effect, Pauli found it "unsightly" and "monstrous." "I am deeply insulted by it," he wrote to Bohr.

Ten days later Pauli wrote to Bohr again, offering his own deeply critical assessment of the situation: "The atomic physicists in Germany can now be divided into two classes. Some work out a given problem first with half quantum numbers, and if it doesn't agree with experiment, they do it again with integral ones. The others calculate first with integral values, and if it doesn't work, do it again with halves." In other words, they had all been reduced to desperate measures.

As far as he was concerned the problem of the anomalous Zeeman effect was far from solved. He was becoming convinced that "there is no [satisfactory] model for the anomalous Zeeman effect and that we have to create something fundamentally new."

But he had no idea what this might be. The whole farrago was getting him down. "I myself have no taste at all for this sort of theoretical physics," he wrote to Bohr, and wanted to withdraw from it. Atomic physics had all become "too difficult."

Dr. Jekyll and Mr. Hyde

Physics was Pauli's heart and soul. His physics research gave definition to his life and his fruitless attempts to solve the anomalous Zeeman effect, on top of what he regarded as his lack of success with the hydrogen-molecule ion and helium atom, began to take a heavy toll on his already fragile psyche. His early successes—his maiden papers on relativity theory—suddenly seemed in the distant past. He began drinking more and more heavily. "I have noticed that wine agrees very well with me," he wrote to a friend. "After the second bottle of wine or champagne I usually adopt the manners of a good companion (which I never have in the sober state) and then may under these circumstances enormously impress the surroundings, particularly if they are women."

By day he behaved like a staid Germanic professor. By night he roamed the Sankt Pauli, Hamburg's notorious red-light district full of risqué cabarets and bars catering largely to a rough clientele. He described his life to a friend: "During the day, calming work, in the night, sexual excitement in the underworld—without feeling, without love, indeed without humanity." Much later, writing to Jung, he recalled "the complete split between my day life and my night life in my relations with women." He seemed to have split into Dr. Jekyll and Mr. Hyde. Robert Louis Stevenson's celebrated novel had been published some forty years earlier. No doubt Pauli had read the story of the scientist who is taken over and destroyed by his darker impulses. Perhaps he saw a parallel between Dr. Jekyll and his own increasingly erratic behavior.

Hamburg was a vibrant city that welcomed all comers and in which one could savor the steamy side of postwar Germany. Munich banned the American cabaret performer Josephine Baker, who was famous for her nude dancing; Hamburg welcomed her with open arms.

The real action was on the side streets off the main Sankt Pauli avenue, particularly on a street called Grosse Freiheit. Even during the day it was difficult to see inside the bars there. The odor of spilled beer and the sticky unmopped floors made the interiors stifling. When Pauli walked in in his fine suit and went to the bar, no doubt in the early days at least conversation would grind to a halt and everyone would stare until he had finished his drink and left. But he soon became a regular. To make things worse, the more he drank, the more obnoxious he became.

Often he ended up getting beaten in a brawl. Once he was eating in one of his favorite restaurants in the area. A row broke out and Pauli found himself right in the middle of it. He only pulled himself together when someone threatened to throw him out of a second-floor window. Afterward, he said, he could not understand how he had gotten into such a situation.

He began to feel as if he were losing control. He was frightened of the person he was becoming. "[I] tended toward being a criminal, a thug (which could have degenerated into my becoming a murderer)," he later recalled. By day, immersed in his research, he felt "detached from the world—a totally unintellectual hermit with outbursts of ecstasy and visions." His two parallel worlds were in danger of colliding with potentially fatal effect.

The women Pauli found in the bars there offered a way to forget his growing frustration and anger. Typical of his Sankt Pauli girlfriends was a beautiful blond woman some two years younger than he. They had a short and passionate affair that Pauli broke off when he discovered she was a morphine addict. Then one day she turned up at his office at the university. Somehow she had found him, despite his secrecy and desperate attempts to keep his night and day lives separate. Pauli was horrified. Poor, sick, and stick thin from her continuing morphine abuse, she stood like a specter, begging him for help. Pauli threw her out, and told her never to come back—and she disappeared back into the Sankt Pauli. He forgot about her, hoping she was gone forever. Little did he guess that in later years she would come back to haunt him.

Pauli always kept his visits to the Sankt Pauli secret, even from his closest friends and colleagues. These included the always upbeat Otto Stern, Emil Artin, Walter Baade, and Gregor Wentzel. Wilhelm Lenz, director of the Institute for Theoretical Physics, joined them from time to time, particularly for departmental lunches, which were always held in top restaurants scouted out by Stern. Like Pauli they were all bachelors.

Lenz was a man of some means who lived in a fashionable area of Hamburg at 18 Armgartstrasse, on a beautiful canal with grassy banks and the city's largest lake, the Aussenalster, glittering in the distance. When Pauli first arrived Lenz offered him a room in his house. Pauli later moved around the corner to 16 Papenhude, where he had an apartment on the second floor. Miraculously the area escaped damage in World War II and remains today much as it was then. Lenz was noted for his reserve.

Otto Stern, an unusually gifted experimental physicist, was another recent addition to Hamburg. Like Lenz, Stern was a rather wealthy man. But unlike Lenz he was outgoing—a bon vivant who sometimes flew to Vienna just for lunch. Artin was a mathematician who specialized in number theory, Baade an astronomer, and Wentzel a physicist. These last three were Pauli's exact contemporaries.

Wentzel was Pauli's closest friend. Not only did their research interests overlap but so did their idea of a good time. Wentzel frequently went to Paris on the slightest pretext. On one occasion he sent Pauli two of his papers to comment on and signed the letter giving his address simply as "Paris." Pauli swiftly replied, "The question is this, whether indicating Paris at the end of your work suffices at least to justify all this psychologically and whether in the corrections you should not change it more specifically into Paris, Moulin-Rouge, or something analogous."

Pauli enjoyed visiting Baade and the astronomers at their observatory in Bergedorf. On full-moon nights it was impossible to observe the stars and they would have a party instead. On one occasion Pauli was present at the observatory when it was discovered that a terrible accident had befallen the great refractor telescope. It was almost destroyed. Naturally everyone chalked it up to the Pauli effect.

Cases of the dreaded Pauli effect were beginning to pile up. Physicists at the university became convinced that Pauli's presence in or even near a laboratory led to severe breakdowns in the equipment. Stern was reduced to desperate measures. He recalled that the only way he could protect his laboratory from the Pauli effect was that Pauli "was not allowed to enter." The Hamburg scientists were surprisingly superstitious. One brought a flower and gave it to his apparatus every day. Stern kept a hammer lying next to his as a veiled threat to it not to break down. Pauli himself fervently believed in the Pauli effect and began to wonder whether he emanated powers.

Pauli's exclusion principle: Four quantum numbers instead of three

Pauli had given up trying to solve the anomalous Zeeman effect, but he kept up with the flood of papers that poured out on the subject. Then in autumn 1924 two ideas suddenly came together for him.

The first was this: Suppose relativity had something to contribute on the subject. Pauli looked into it. He found that in the models of the atom proposed by Bohr and Heisenberg, electrons within the core moved at speeds comparable to that of light. They should therefore be expected to display variations in mass consistent with Einstein's equation $E = mc^2$. These variations should show up in the spacing between the multiplets, but experiments had not revealed any such effect and could mean only one thing: the core in these models had to be inert; it did not interact and so played no role at all. In other words, every model of the atom that featured a core was wrong.

Then he came across a paper by Edmund Stoner, a twenty-five-year-old physicist at Leeds University. Stoner went far beyond the anomalous Zeeman effect, although he himself had not realized the full significance of what he had found. It had to do with the problem constantly on Pauli's mind: What stopped every electron in an atom from falling into the atom's lowest energy level—its ground state?

By clever manipulation of the three quantum numbers for an electron in an atom, Stoner had succeeded in calculating the total number of multiplets of an alkali atom undergoing the anomalous Zeeman effect (that is, when it is placed within a weak magnetic field). He did this, as Heisenberg and Bohr had, by imagining the alkali atom to be made up of a closed core—made of shells filled to their maximum with electrons and so inactive chemically—with a single lone electron revolving around it. From this he was able to show that the total number of electrons in each closed shell was related to twice the total amount of angular momentum of the closed core with the lone electron.

What struck Pauli was the appearance of the number two. Bohr had inserted this number into his model of the core simply so that only halves would appear in formulas for the anomalous Zeeman effect. In other words, when the atom was in a magnetic field, the core containing the closed shells full of electrons could be distorted in two ways, which would give one of its quantum numbers a value of plus or minus a half.

But Pauli had established that the core was inert and that only the lone electron played any role in the chemical activity of an alkali atom. So why not transfer the two possible values of the core to this electron? Pauli began to suspect that Stoner's work contained the seeds of something new and exciting. He decided to see what would happen if he extended Stoner's method of manipulating quantum numbers to include a fourth quantum number that had the values of plus and minus a half for the lone electron. The result was astounding. He figured out that the total number of electrons in each closed shell was twice the principal quantum number of that shell squared. It was $2n^2$, the same number that Bohr had proposed with no basis from his theory of the atom. Now there was one.

Pauli went yet further, proposing that the two possible values for the fourth quantum number be assigned to every electron in every atom, regardless of whether the atom was in a magnetic field.

The conclusion had to be that each electron in an atom required four not three quantum numbers, and, to explain the periodic table of chemical elements, that no two electrons in an atom could have the same four quantum numbers. Basically, two electrons with the same quantum numbers cannot occupy the same shell. (This is Pauli's famous exclusion principle. The name was given it by Paul Dirac, a physicist at Cambridge University.) This was the reason why Bohr's building-up principle for atoms worked—why there are precisely two electrons in the inner shell, eight in the next, then eighteen, and so on. This was also why every electron in an atom did not fall into its lowest stationary state. They were prohibited from doing so.

To get a grip on this complicated concept, imagine that an atom is an apartment building with many rooms on many different floors and no elevators. In fact, it is an upside-down pyramid with two rooms at the bottom, eight on the second floor, eighteen on the third, and so on. To avoid overcrowding, the local housing authority passes a law that only one electron can occupy a room. A crowd of electrons enters the building and jostles around, trying to occupy as low a floor as possible. No electron wants to be the only one on a floor—the lone electron, as in an alkali atom. Such an electron cannot relax because on its shoulders rests the chemical activity for the entire atom. This is a very simplified description of the way that Pauli's exclusion principle works.

Pauli's paper on the exclusion principle contained none of the mathematical fireworks for which he had become famous. Rather, it was the fruit of his patient examination of data. By searching out patterns among numbers, he came up with what scientists call a restrictive or prohibitive principle. Another example is the principle of relativity, which asserts that the laws of physics must be the same in every laboratory, regardless of its motion. There is no reason for this to be so. Yet it must be, to formulate a systematic theory of how objects from the size of basketballs to planets move. It also enables scientists to predict numerous phenomena that no one had previously thought of, such as the bending of starlight by massive objects, a prediction of relativity theory that was later proved to be true in real life. There is no way of deriving the principle of relativity mathematically. It is simply an axiom.

But what about the exclusion principle? Could it be derived? Pauli was not sure, nor was anyone else.

Hard though it was to understand its deeper meaning, scientists quickly realized the exclusion principle's importance in explaining the periodic table of chemical elements and thus, also, atomic structure. It also helped clarify why metals are hard and what the fate of stars might be. Pauli had made a discovery that would shape the path of physics in the future and change our understanding of the cosmos.

The search for the meaning of the exclusion principle

Pauli immediately notified Bohr and Heisenberg of his discovery and sent them the draft manuscript of his paper. It was, he wrote them firmly, at the very least *"not a bigger* nonsense" than the schemes other scientists proposed for understanding the structure of the atom. At least Pauli had avoided hypotheses with no basis such as Bohr's force of unknown origin, which distorted a core in two different ways. He suspected that his exclusion principle could not be derived from Bohr's theory of the atom. Understanding it lay rather, he suggested, in the as-yet-unknown properties of "motion and force in quantum theory." Remembering his many attempts to support Bohr's theory—the hydrogen-molecule ion, the helium atom, and the core models of atomic structure—all of which ended in failure, he wrote that he would prefer to interpret the exclusion principle free of any model of the atom, especially a model containing the concept of electron orbits. He was sure that the key factors in describing the characteristics of an electron had to be its energy and momentum. Those were real because they were measurable; electron orbits and shells were not. In this Pauli was true to his godfather Ernst Mach's philosophy—to avoid any unmeasurable concepts in a theory of physics, for those were purely metaphysical.

Heisenberg and Bohr were amused by Pauli's exclusion principle. Here was proof, Heisenberg wrote to Pauli, that Pauli was entering the "land of the formalist philistines," practicing a style of physics "of which you had insulted me. [In fact, you] had broken all hitherto existing records [in rising] to an unimagined, giddy height (by introducing *individual* electrons with 4 degrees of freedom)." Everyone knew that electrons had to move in three-dimensional space like everything else in the universe, and therefore three quantum numbers should surely suffice. Pauli had frequently accused Bohr and Heisenberg of coming up with "swindles." Now Heisenberg accused Pauli of coming up with a swindle of his own; "swindle x swindle does not yield something correct," he wrote.

A week later, Bohr had clearly thought more seriously about Pauli's new theory. He wrote to him, "I have the impression that we stand at a decisive turning point, now that the extent of the whole swindle has been so exhaustively characterized." What struck him about Pauli's proposal, he said, was its "complete insanity." Bohr always condemned new proposals with the words "interesting but not crazy enough." Saying that Pauli's was completely insane meant he thought it was most probably right.

Pauli had still not solved the anomalous Zeeman effect, but he had accomplished something far more important. He suspected that the full significance of the exclusion principle would not become clear until there was a deeper understanding of quantum theory. "I will wait patiently; and be satisfied if I live to see the solution," he wrote to Bohr. He hoped his "insane idea" would help toward understanding the structure of atoms made up of many electrons. If it did, "I would be the happiest man on earth."

The fourth quantum number

The problem people had with understanding the exclusion principle was the lack of a visual model for the fourth quantum number. Pauli was well aware that it was essential for physicists to be able to visualize a theory—which was what made Bohr's image of an atom as a miniscule solar system so pleasing. Nevertheless, he wrote to Bohr, although this need for visual images was "in part legitimate and healthy, it should never count as an argument for retaining systems of concepts. Once the systems of concepts are settled, then will visualizability be regained." In effect, Pauli was suggesting strongly to Bohr that he should drop all visual images from his theory because they had proved to be incorrect and misleading. It was only once a new theory of atomic physics had emerged that it would be possible to develop a visual language to describe the atomic world.

Then Ralph Kronig, a twenty-year-old German American, noticed something Pauli had overlooked. The fourth quantum number of an electron has the mathematical properties of angular momentum, the momentum of an object moving in a circle. Every electron has an angular momentum from its orbit around the nucleus of the atom, like the earth revolving around the sun. Perhaps, thought Kronig, each electron also has an angular momentum of its own, like the earth spinning on its axis.

But whereas the earth's spin is variable, the electron's always remains the same. Kronig gave electron "spin" a value of a half, using the units of angular momentum. To explain spectral lines physicists had always assumed that the electron acted like a tiny magnet that could align itself in a magnetic field. Pauli's and Kronig's discovery had transformed it into a spinning magnet that could align itself along a magnetic field in one of two directions, depending on whether it had a spin of plus or minus a half. These were precisely the two values that Pauli had transferred from the inert closed core to the lone electron in the outermost shell of an alkali atom. Thus spin was recognized as a distinguishing feature of an electron. Every electron has its own spin, just as each of us has a nose, eyes, and lips that distinguish us from one another. Spin is an intrinsic property of an electron, and no matter where the electron is located it has a spin of a half.

Initially Pauli dismissed Kronig's proposal, saying merely that it was "indeed a witty idea." For imagining an electron as a spinning top, as Pauli and everyone else did, led to a serious conflict with relativity theory. It meant that a point on the surface of the electron might move with the velocity of light, which according to relativity theory is impossible. When Kronig visited Copenhagen, Bohr dismissed his proposal with the words "very interesting"; Kronig dropped the idea.

Then, nine months later, two Dutch physicists, George Uhlenbeck and Samuel Goudsmit, rediscovered spin and staked their claim in print, warning that one should not visualize the electron as a spinning top. Pauli was deeply embarrassed at having discouraged Kronig from publishing his idea and thereafter always spoke highly of him.

Spin was undeniably a property of an electron but it was entirely impossible to visualize it in a way consistent with relativity theory. Scientists had to accept that the fourth quantum number had no accompanying visual image. It was time for atomic physics to move on from trying to visualize everything in images relating to the world in which we live.

5

Intermezzo—Three versus Four: Alchemy, Mysticism, and the Dawn of Modern Science

My branch of science, physics, has got somewhat bogged down.
The same thing can be said in a different way: When rational
methods in science reach a dead end, a new lease on life is given
to those contents that were pushed out of time consciousness in
the 17th century and sank into the unconscious.
—WOLFGANG PAULI

SINCE his student days and before, Pauli had been interested not just in the rational world of physics but also in the role of the irrational. Arnold Sommerfeld, his professor and lifelong mentor, was fascinated by the sixteenth-century pioneer of modern science, Johannes Kepler. Science, Sommerfeld reminded his students, emerged out of mysticism and had never completely separated itself. Besides his purely scientific work, Sommerfeld also pursued kabbalistic lines of research based on pure numbers and spoke of Kepler as his precursor.

In fact, he saw a direct connection between the developments of modern science and the harmony that Kepler had been searching for. Writing of his own research into spectral lines, the "fingerprint" of the atom, he said, "What we are nowadays hearing of the language of the spectra is a true music of the spheres within the atom, chords of integral relationships, an order and harmony that becomes even more perfect in spite of manifold variety."

Pauli, too, sought out links between his work and these ancient esoteric systems of understanding the universe. Many years later he was to write to Sommerfeld that he had finally succeeded in using the exclusion principle to establish the reason why the electron shells fill up as they do in the series 2, 8, 18, 32, a grouping that Sommerfeld had described as "somewhat kabbalistic." In 1923, when Sommerfeld wrote those words, no one had been able to find any reason why electron shells should fill up in this way and not some other. There was, in fact, no firm basis for almost all of the rules of atomic physics. "Some of the rules recall irresistibly the teaching of the alchemists or the witches' kitchen of Faust," wrote one physicist. Writings on the subject were nearly as mysterious as the Kabbalah, the Jewish book of mysticism that claimed that numbers could yield insight into the world beyond sense perceptions, just as Bohr's theory of the atom promised.

From time to time Pauli could not resist poking fun at his mentor's obsession with numbers. Once he noticed an advertisement posted around Munich, promoting an optical firm: "If you have trouble with your eyes, see Herr Runke." Pauli added the coda, "For integers, go to Sommerfeld."

In a tribute to Sommerfeld on his eightieth birthday, Pauli wrote that he "would not hesitate to set as a superscription over Sommerfeld's works in a wider sense the title of Kepler's magnum opus—*Harmonices Mundi.*"

Inspired by Sommerfeld, Pauli became fascinated by Kepler and may well have read him in the original Latin. In Hamburg he could have followed up his interest by attending lectures by Erwin Panofsky. A historian of art who specialized in symbolism and iconography, Panofsky had a great deal of knowledge about early science, particularly Kepler.

But what was it about Kepler that particularly piqued Pauli's interest?

It seems likely that Pauli read Kepler's *Harmonices Mundi* (Harmonics of the World) at this early stage of his life, when he was working on the exclusion principle. If so, he could not fail to have noticed the Appendix to Book V, about Robert Fludd, with whom Kepler had clashed. Kepler had stood up for the number three as the key number necessary to explain the workings of the universe. Fludd, conversely, asserted that it was four. "I, myself, am not only Kepler, but also Fludd," Pauli was to write many years later. For Pauli had been in exactly the same quandry.

"Once (in Hamburg) my path to the Exclusion Principle had to do precisely with the difficult transition from 3 to 4, namely with the necessity to attribute to the electron a fourth degree of freedom (soon explained as "spin") beyond the three translations. . . . That was really the *main work*." Bohr and Heisenberg had asserted firmly that there could only be three quantum numbers for an electron. But Pauli, almost despite himself, realized there had to be four.

A few years later he was to come across the same numbers again in Carl Jung's psychology, based as it was in alchemy. He was adamant that "in neither case was it by any means Mr. C. G. Jung who suggested it to me, nor was there an advance conscious intention for me to grapple with figuring out the problem of three and four. Consequently I am rather certain that *objectively* there is an important psychological and, perhaps, natural philosophical problem connected with these numbers."

Pauli's study of Kepler and Fludd

Pauli's study of Kepler and Fludd, "The Influence of Archetypal Ideas on the Scientific Theories of Kepler," published in 1952, is the definitive work on the two and indicative, incidentally, of Pauli's creative ability as a historian of science. His goal, as he wrote, was not so much to enumerate facts as to explore the "origin of and development of concepts and theories." Scientists such as Einstein and Poincaré had insisted on the importance of intuition in creative thinking. Logic alone cannot lead to the discovery of scientific theories, they said. But neither man suggested a way to bridge the gap between intuition and the precise concepts required of a scientific theory.

Pauli's inspiration was to look into the practice of science in the Middle Ages, when alchemy, astrology, myths, the Kabbalah, and magical symbolism were all accepted modes of thought. It was a time of enormous change, when thinkers were daring to question the authority of the Church in understanding nature and the process of learning itself was becoming secularized. As scholars debated what questions they should ask about the world around them, brilliant men were developing the fundamental methods of science. Gradually a strongly rationalistic, logical science developed, forcing the more irrational, mystical elements into the background, where perhaps they remained in the unconscious of modern scientists.

Kepler was sure there was an order and harmony in the world and struggled to find a means to comprehend it. Straddling two worlds, he suffered great personal anguish in his quest to find the proper place for God. Again and again his research came up with precisely the symmetries and harmonies he was expecting, yet it always seemed to result in a universe without a God. Robert Fludd, meanwhile, remained firmly entrenched in the Middle Ages and in the end the two had to clash. It was a collision between two opposing intellectual worlds, leading each of them to produce a one-sided and incomplete understanding of nature. The forces of mysticism were at the time still overwhelming while science as we know it today was still in its infancy—but a beacon of light.

Early travails

Johannes Kepler was born on May 16, 1571, in the town of Weil-der-Stadt in Germany. His father was a swashbuckling soldier of fortune who fought for anyone willing to pay for his services, even the Catholics, which disgraced the Keplers in the eyes of the Protestant families of Weil. He vanished in the course of a mercenary adventure. Kepler's mother, Katharina, was said to be quarrelsome and generally unpleasant, just like her husband.

A sickly child, Johannes grew up in what he described as a virtual madhouse surrounded by his squabbling parents and grandparents and six siblings, one of whom suffered from epilepsy. At four he almost died of smallpox, and two years later his hands were badly crippled. At sixteen, he wrote, he "suffered continually from skin ailments, often severe sores, often from the scabs of chronic putrid wounds in my feet which healed badly and kept breaking out again. On the middle finger of my right hand I had a worm, on the left a large sore." At twenty-one he was "offered union with a virgin; on New Year's Eve I achieved this with the greatest possible difficulty, experiencing the most acute pains of the bladder."

Given his poor health, interest in religion, and excellent record in the elementary Latin school, the obvious choice of career was to join the clergy. In 1589 he entered the theology school at the University of

Tübingen. The professor of mathematics and astronomy there, Michael Maestlin, invited him to join his private study group.

Copernicus's sun-centered universe

In his public lectures Maestlin taught Ptolemy's model of the universe with the sun and stars circling the earth, which agreed with Christian theology. But in private he was intrigued by the world-picture proposed by the Polish priest Nicholas Copernicus in his book of 1543, *De Revolutionibus* (On the Revolutions), which put the sun, not the earth, at the center of the planetary orbits.

In Copernicus's model, the six planets (Mercury, Venus, Earth, Mars, Jupiter, and Saturn) make circular orbits around the sun, all enclosed by the sphere of the fixed stars. But Copernicus was perfectly aware that

*Copernicus's 1543 model of the universe. The sun (*sol*) is at the center with the six planets moving around in circular orbits. All this occurs within the seventh sphere—the sphere of the fixed stars (*Stellarum fixarum*). (Copernicus, De Revolutionibus [1543].)*

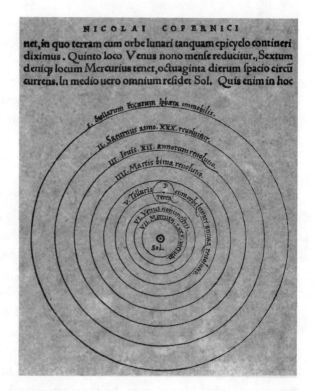

planets do not move in this way. A planet moving from west to east might cut back to the west, then resume an easterly orbit (known as retrograde motion). To explain these more complicated motions, Copernicus set planets moving on circles whose centers were on the surfaces of other circles that were also in circular motion, making the motions of planets the sum total of many circular motions. When more precise observations resulted in one of these planetary systems falling out of line, he added yet more circles. Thus he was able to explain the observed motion of the planets.

Church authorities condemned Copernicus's system as heresy, in that even though his system offered the best available mathematical basis for a twelve-month calendar, it did not place the earth at the center of the universe. (By "universe," Copernicus and his contemporaries meant what we now know as the solar system.) Church scientists expurgated *De Revolutionibus*, declaring that assertions made with certainty were in fact merely hypothetical. But Maestlin believed otherwise: that this was the way things actually were.

Kepler, too, read Copernicus's book. Two passages fired his imagination. In one Copernicus wrote, "In the middle of all sits the Sun enthroned. In his beautiful temple could we place this luminary in any better position from which he can illuminate the whole at once? He is rightly called the Lamp, the Mind, the Ruler of the Universe; Hermes Trismegistus names him the Visible God, Sophocles' Electra calls him the All-seeing. So the sun sits as upon a royal throne ruling his children, the planets which circle around him."

In the other Copernicus wrote of his model of the sun-centered universe: "We find in this arrangement a marvelous symmetry of the world and a harmony in the relationship of the motion and size of the orbits, such as one cannot find elsewhere." These words struck Kepler like a bolt of lightning. He was gripped by the notions of order and harmony.

More than two hundred years were to pass before scientists were able to give incontrovertible proof that the earth circled the sun by measuring stellar parallax—the change in a star's position caused by the earth's movement around the sun. The only proof Copernicus had was his sense of aesthetics and harmony phrased in the language of mysticism. But that was good enough for Kepler.

"Geometry is the archetype of the beauty of the world"

Kepler believed in a reality beyond appearances. To him the three-dimensional sphere was the most beautiful image because it symbolized the Holy Trinity, the Triune God, with God the Father at the center, the Son at the circumference, and the Holy Ghost emanating from the center, as the radius. Thus there was an unchanging relationship between the circumference, the radius, and the center point. As he put it, "Although Center, Surface, and Distance are manifestly Three, yet are they One." The curved surface of the sphere with no beginning and no end represented the eternal Being of God.

The universe, with the sun at the center and the planets revolving around it in three-dimensional space, was the perfect sphere and thus the very image of the Holy Trinity. Kepler saw it as a triumph of geometry, the discipline which to him ranked highest among the sciences. Pauli quotes from Kepler three different assertions of this:

> The traces of geometry are expressed in the world so that geometry is, so to speak, a kind of archetype of the world.

> The geometrical—that is to say, quantitative—figures are rational entities. Reason is eternal. Therefore the geometrical figures are eternal; and in the Mind of God it has been true from eternity that, for example, the square of the side of a square equals half the square of the diagonal. Therefore, the quantities are the archetype of the world.

> The Mind of God, whose copy is here [on earth] the human mind, from its archetype retains the imprint of the geometrical data from the very beginnings of mankind.

Writing in 1952, many years after he began his association with Jung, Pauli could not have failed to notice the appearance again and again of the word "archetype." The axioms of geometry, Kepler believed, are imprinted in our minds from birth by the Supreme Geometer. "Geometry is the archetype of the beauty of the world," he wrote.

In his thinking Kepler was influenced by the philosopher Proclus, who lived in the fifth century A.D. Proclus believed that mathematics—to be specific, whole numbers—held the key to understanding the nature of God, the soul, and the world-soul—the ethereal order and beauty of the cosmos. He wrote of an eternal unchanging universe, governed by laws of mathematical order and different from the imperfect world in which we live. His universe emanated from the One. Later commentators interpreted this One as a fecund Deity who gives light, warmth, and fertility.

Echoing Proclus, Kepler argued that it made sense to assume that the sun, not the earth, was at the center of the universe. Only thus could it "diffuse itself perpetually and uniformly throughout the universe. All other beings that share in light imitate the sun." The sun, its light, and the sphere of the fixed stars reveal the Holy Trinity before our very eyes. As Pauli put it: "*because [Kepler] looks at the sun and the planets with this archetypal image in the background he believes with religious fervour in the heliocentric* [sun-centred] *system. . . .* [It is his religious belief that impels] him to search for the true laws of planetary motion."

Thus the discovery of the sun-centered universe and the mathematical concepts that went along with it, including three-dimensional space, could be traced to the visual image of the abstract sphere representing God the Trinity—an archetype from deep in the collective unconscious. This was a product of a geometry, wrote Kepler, that "supplied God with the models for the creation of the universe." The sun-centered universe reflected God's glory and this was why Kepler was impelled to search out its laws.

Pythagoras, apostle of fourness

Pauli traced the origins of Kepler's thinking back to the Greek scientist and priest Pythagoras, who lived around 500 B.C.

It was Pythagoras who pioneered the quest for a link between numbers and the cosmos. Pondering the hidden meanings of the world around him as he played on his lyre he began to wonder whether the laws of harmony depended on numbers. He found that to play tones an octave apart, the length of the strings needed to be in specific ratios: the keynote of the octave sounded when the ratio was 1:2; a fifth required a ratio of 2:3; and a fourth, 3:4.

Perhaps numbers might belong to a world beyond perception, which could only be fully apprehended by thought. His striking conclusion was that the numerals 1, 2, 3, and 4 represented all known objects: 1 represents a point; 2 points can be connected by a line; 3 points make a triangle, in particular a perfect equilateral triangle; and 4 points make a tetrahedron, a pyramid of three perfect triangles. From these could be constructed the five "Pythagorean" solids (later "Platonic" solids after Plato): the tetrahedron, cube, octahedron (eight equilateral triangles), dodecahedron (twelve pentagons), and icosahedron (twenty equilateral triangles). Each could be circumscribed by a sphere, with each point of the solid touching its surface, and each could also contain a sphere whose surface touched each of its sides.

Represented as dots, 1, 2, 3, and 4 form an equilateral triangle set out in four rows, known as the *tetraktys* (*tetras* is Greek for "four"):

Pythagoras's tetraktys.

To Pythagoras this analysis made sense of our world, in which he recognized four elements (earth, water, air, and fire), four seasons, four points of the compass, and four rivers of paradise (the Pishon, the Gihon, the Tigris, and the Euphrates). His followers swore an oath "by him who has committed to our soul the tetraktys, the original source and root of eternal Nature." The sum of the numbers that made up the tetraktys is ten, which Pythagoras considered the perfect number. Once we have counted to ten, we return to one, the number of creation.

Pythagoras's claim was that numbers were the fabric of our universe and existed independently of us. Numbers were the keys through which could be heard the harmony of the cosmos.

The Kabbalah

The Egyptian god Thoth, known to the Greeks as Hermes Trismegistus ("Thrice-great Hermes"), was credited with a huge number of writings on philosophy, astrology, and magic. Over the centuries Hermetic literature incorporated elements of whatever science existed as well as the teachings of Pythagoras.

In Kepler's time Hermetic literature was enthusiastically embraced as an antidote to the rational approach of Greek philosophy and science. It was full of mystery and magic and spoke in terms of a vital or living force at the heart of the cosmos. Hermetic literature also included kabbalistic texts.

Versions of the Kabbalah had begun to appear in the thirteenth century. A principal theme was how one might see the invisible in the visible and the spiritual in matter. The Kabbalah discussed the clash between opposites like light and darkness to produce the world in which we live. Someone like Kepler, who was interested in the teachings of Proclus, was naturally drawn to the Kabbalah with its similar theme.

A central notion of kabbalistic philosophy is the *Sephirot*. The Sephirot is usually represented as the tree of life with ten branches rooted in the earth and extending to Heaven, signifying the earth as a microcosm reflecting the universe, the macrocosm. It is made up of five pairs of opposites—beginning and end, good and evil, above and below, east and west, and north and south—and thus has ten emanations, ten being a holy number in Judaism as well as in Pythagoreanism.

By the end of the fifteenth century the Kabbalah had been integrated into Christian theology, though the Christian Kabbalah emphasized the Trinity rather than the Sephirot. Christian thinkers were especially fascinated by the *Gematria*, which assigned numbers to letters of the Hebrew alphabet. This concept of numbers for language opened up the possibility of assigning numbers to the various names of God, thereby further revealing His celestial powers and His mystery. Thus the Kabbalah became identified with magic and numerology. (Until the nineteenth century the Hebrew alphabet had no numbers; letters were used for numbers. Thus in Roman times 666 happened to be the letters for Nero's name.)

By Kepler's day the Christian Kabbalah was considered one of the "handmaidens" of true wisdom, along with alchemy and astrology. But all this clashed with the onset of a new, materialistic science that claimed to be able to predict the course of cannonballs and planets using mathematics, but only if a division were made in nature between dead and live matter. For mathematics could be applied only to the former, not to the latter.

Kepler's model of the universe

When Kepler was growing up, there was a flood of astrological, kabbalistic, and alchemical texts being published. Anything attributed to Hermes Trismegistus was hailed as a revelation. They held readers spellbound, the vaguer the better. Kepler was hooked; his enormous imagination was sparked.

Why was the world as it was? Why were there six planets (Mercury, Venus, Earth, Mars, Jupiter, and Saturn)? Why were they at certain distances from the sun? What was the relationship between their distances from the sun and their speeds? Might the answers to these questions lie in certain arrangements of geometrical figures?

By this time Kepler was a district mathematician, teaching mathematics and astronomy at the Protestant Seminary in Graz. During one of his classroom lectures on geometry he happened to draw an equilateral triangle. Inside it he drew a circle touching all three sides and around it another circle touching its points, just as Pythagoras had described. Suddenly it all fell into place. It was a model of the universe.

Clearly the reason why there were six planets was because there were five perfect solids symbolizing the five intervals between the planets. The planets moved on spheres that circumscribed the five solids. Calculating the distances of the planets from the sun, Kepler drew up a new image of Copernicus's universe with the sun at the center and nested planetary spheres on which the planets moved. The sphere of Mercury inscribed an octahedron, that of Venus an icosahedron, that of Earth a dodecadron, that of Mars a cube, and that of Saturn a tetrahedron, which the orbit of Jupiter circumscribed.

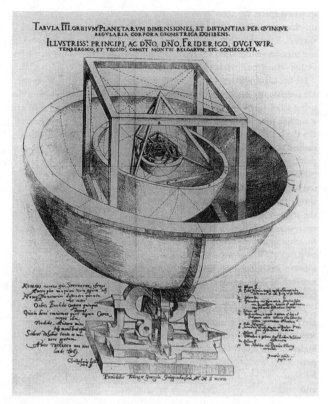

Kepler's 1596 model of the universe. (Kepler, Mysterium
Cosmographicum *[1596].)*

Kepler attributed this revelation to "divine ordinance." He had
"always prayed to God [that] Copernicus had told the truth." In his diary,
he noted the fateful day when God spoke to him: July 19, 1595.

He was convinced he had discovered God's geometrical plan of the
cosmos, which God had made in his own image. He published his work
in 1596 in a book entitled *Mysterium Cosmographicum* (The Mystery of
the Universe).

However, his model was not in total agreement with Copernicus's
data, especially the data for the orbit of Mercury. Despite his mystical
leanings, Kepler was a new breed of scientist. He required theories to be
supported by data. He decided that what he needed was more precise
data. Copernicus's were not good enough.

From circles to ellipses

Among the people to whom Kepler sent his new book was the greatest observational astronomer of the day, Tycho Brahe. Kepler by now was a handsome twenty-five-year-old with a high forehead, immaculate goatee, aquiline nose, and a look of piercing intelligence. Tycho, as he was always known, was twice Kepler's age. He sported a mustache so immense that it looked like a walrus's tusks and was famous for his prosthetic nose, having had his real one cut off in a duel. He had a copper nose for everyday and a gold and silver one for special occasions.

Tycho achieved his world-renowned accuracy by making all his observations from his monstrous-looking observatory, Uraniburg, on the island of Hveen, off the coast of Denmark. But to Kepler's annoyance, Tycho refused to reveal his data to anyone until he had refined his own model of the universe—in which every planet except the earth orbited the sun and the entire assemblage, in turn, orbited the earth.

Impressed with Kepler, Tycho offered him a position as his assistant so that he could help him with the mathematics of his model, little realizing that Kepler simply wanted to lay his hands on his data. Kepler accepted

Johannes Kepler.

Tycho Brahe.

the offer and joined Tycho in Prague, where he was imperial mathematician to the court of Emperor Rudolf II.

Tycho set Kepler to work to improve his observations of Mars, the most difficult of the planets due to its pronounced retrograde motion. Astronomers described the orbit of Mars as having a large "eccentricity"—the distance that the sun had to be moved from the center of Mars's orbit to improve agreement with Tycho's data of the complicated system of circles rolling on circles. This displacement was a mathematical device used in every model of the universe—in Ptolemy's it was the earth whose position was displaced from the center of the universe. In reality, of course, in Copernicus's system, the sun was at the center of the universe. The models of Ptolemy, Copernicus, and Tycho could not deal adequately with Mars's eccentricity. Kepler bet his colleagues that he could straighten it out in eight days. In fact it took him eight years.

In October 1601 Tycho suddenly died at the age of fifty-five and Kepler was appointed imperial mathematician. He inherited all of Tycho's data and more important, no longer had to waste time fiddling with Tycho's model of the universe.

To start with, Kepler analyzed Tycho's data on the orbit of Mars, trying to preserve the old model of the universe by explaining the orbits of the planets in terms of circles. Taking Tycho's best data for Mars, he used the mathematical device of displacing the sun from the center of Mars's orbit by a certain distance to allow for eccentricity. Then, by adroit mathematics, he moved himself from the earth to Mars and found that the earth also moved in an orbit similar to Mars's, with varying speeds.

Supposing the orbit of Mars was not a circle but an oval? Kepler spent 1604 struggling with the mathematics of an oval. That year was full of problems. Both he and his wife fell ill; when he became short of money his wealthy wife refused to dip into her funds; and she also gave birth to yet another child whom Kepler saw as yet another problem. And an ominous new star appeared in the sky—the nova of 1604.

Then he tried replacing the oval with an ellipse. An ellipse is a circle that has been squashed at its north and south poles. It has two centers, or focii, neither of which is in the middle. When the two centers are moved together the ellipse becomes a circle. This worked perfectly. The curve went through all of Tycho's data points for the orbit of Mars and also fitted Mars's measured eccentricity. Kepler had discovered his first law of planetary motion: that every planet moves in an ellipse with the sun at one of its centers. The sun is no longer at the center of the universe but at one of the ellipse's foci.

Soon after, he discovered his second law: that a line drawn from the sun to a planet sweeps out equal areas in equal times. This meant that a planet's speed varied as it traveled in an ellipse around the sun: the planet sped up as it neared the sun and slowed down as it moved away.

Kepler had overthrown the two-thousand-year-old assumptions that the complicated orbits of planets could only be explained by adding circles moving on circles in uniform circular motion and that the planets move with a uniform speed. He published his new laws of astronomy in his 1609 *Astronomia nova* (The New Astronomy).

But what kept the planets from escaping altogether and flying off into the void? Perhaps there were tentacles emanating from the sun, grasping a planet and whipping it around in its orbit. Kepler imagined the attraction to be magnetism. Newton would later discover that it was gravity. Kepler, however, could only conceive of it as some sort of vital or living force.

As Pauli points out, Kepler was caught between two worlds. His laws of planetary motion were an accurate description of the paths of the planets around the sun, but they emerged from mathematical calculations that wrenched the sun out of its true place at the center of the universe. Using mathematics meant he had to treat the earth as dead matter. However, according to his Renaissance beliefs the earth was not dead at all, it had a soul, an *anima terrae*, akin to the human soul. It was a living organism. Sulphur and volcanic products were its excrement, springs coming from mountains its urine, metals and rainwater its blood and sweat, and sea water its nourishment. Kepler's attempts to link such animistic beliefs to scientific data made him a new breed—a scientific alchemist. He had no choice but to compartmentalize his work: ellipses were confined to the scientific side of his life, circles and spheres to the religious and alchemical side.

Kepler's third law

In 1611 Emperor Rudolf abdicated. To escape the dangerous political intrigues that followed, Kepler moved to Linz, the capital of Upper Austria, a charming city on the Danube. Before leaving Prague, however, his wife fell ill with typhus and died.

Kepler's marriage had not been happy. Nevertheless, after her death he was lonely. He also had three young children to look after, two girls and a boy. He looked around for another wife in the same way he had discovered his two laws—by trial and error. He ended up with eleven choices, some of whom he had advertised for, others whom he had tried out, sometimes boarding his children with them to see if they all got along. One was attractive but too young, another fat, another was of poor health. Kepler finally settled on number five and she gave him the peace of mind to resume his scientific research.

His first two laws had been essentially geometrical—number was missing. Now he turned his attention to numbers. If the sun controlled the planets, he thought there had to be a relationship between the planets' distances from the sun and their speeds.

Meanwhile Europe was heading for the Thirty Years' War. Troops were on the move causing famine, havoc, and plague. Then one of his daughters died. In his grief he turned inward to "contemplation of the *Harmony*," which he believed to exist in nature. Thinking of the musical harmonies explored by Pythagoras, he pondered the eternal reality of numbers, which revealed the very essence of the soul.

How did this numerical harmony relate to the planets in a sun-centered system? Kepler tried to find a way to work out whether harmonious ratios could be formed out of the planets' periods of revolution, their volumes, their sizes, or their velocities when they were furthest from and closest to the sun. But he failed. Then he thought of examining the ratios of a planet's angular velocities at its extremes from the sun, that is, its change in angle at any period as it moved across the sky. And finally the astral music of the Divine Composer began to emerge.

Little by little Kepler worked out the ratios that produced the melodies played by the planets as they moved in their elliptical orbits. It was a heavenly symphony "perceived by the intellect, not by the ear," he wrote. But for Kepler it was much more. To him the planets sang in "imitation of God" in different voices—soprano, contralto, tenor, and bass. But on the earth there was only discord: "The Earth sings Mi-Fa-Mi, so we can gather even from this that *Mi*sery and *Fa*mine reign on our planet," he wrote despondently.

Kepler's third law asserts that the following two quantities are proportional: the time needed for the earth to go once around the sun, multiplied by itself (that is, squared); and the earth's average distance from the sun, multiplied by itself three times (that is, cubed). It completed for him what had been the goal of Pythagoras: to explain the universe in terms of geometry *and* number. He scoured tables of numbers until he found the pattern but he never revealed precisely how he had discovered this capstone of his life's work. He recorded the date: March 8, 1618. "At first I thought I was dreaming," he wrote in the book he published the following year, *Harmonices Mundi*.

Sure that God had spoken through him, he wrote that he did not mind if his book had to "wait a hundred years for a reader. Did not God wait six thousand years for one to contemplate His works?"

All this took place at a time of great personal difficulty. Kepler's mother, Katharina, had been put on trial for witchcraft. Her sister had been burned at the stake as a witch, and this, together with her husband's disappearance, rendered her very suspicious to the gullible populace. In old age she was far from lovely and had a nasty temperament that made her an easy target in the witch-hunting mania in Germany of the early seventeenth century, so much so that she came close to sharing her sister's fate. In 1615, she was in the middle of a feud with another old woman. This neighbor persuaded an influential relative to accuse Katharina of making her extremely ill by feeding her a witch's potion. Others soon began to remember becoming seriously ill after having accepted drinks from Katharina.

Not only was his mother in danger, but so was the family name. Kepler had to take time off from pondering the universe to defend his mother for whom he felt affection and pity, despite his horrendous childhood. The proceedings took over six years. At one point jailers flourished instruments of torture and execution in front of Katharina's face, as was

Robert Fludd.

customary. Unusually for the time, the story has a happy ending. Kepler finally succeeded in obtaining her release.

Robert Fludd—a universe made up of fours

Two years before he finished *Harmonices*, in 1617, Kepler happened to see a highly illustrated book at the Frankfurt book fair: *A Metaphysical, Physical and Technical History of the Macro- and the Micro-Cosm*, by the English physician and Rosicrucian, Robert Fludd.

While Kepler's family was low class, Fludd's was noble. His father, Thomas Fludd, had been knighted by Queen Elizabeth I for his services as war treasurer in the Netherlands and paymaster to English troops in Provence. In portraits Fludd looks rather plump and well fed, with a pointed goatee. He holds his two middle fingers pressed together, perhaps in some sort of secret sign. In one portrait he has fingernails as long as a mandarin's.

Fludd studied at Oxford and became intrigued by Greek philosophy. As was the custom for wealthy young gentlemen, he toured France and Germany, meeting and sometimes tutoring nobility. In Germany he became acquainted with a secret society who called themselves Brothers of the Rosy Cross—Rosicrucians. They called for a reform of knowl-edge in preparation for Armageddon and claimed access to deep secrets and truths in medicine, philosophy, and science. Governments deemed their mysticism and apocalyptic message dangerous and they were often charged with heresy and religious innovation, serious offenses in those days.

When Fludd's enemies at the court of King James I accused him of collaborating with them, he argued persuasively that the Rosicrucians were innocent of heresy. James was so impressed that he became Fludd's patron.

In his book Fludd asserted that "the true philosophy . . . will suf-ficiently explore, examine and depict Man, who is unique, by means of pictures." In other words, he intended the sumptuous illustrations in his books not merely as decoration but as saying something very definite about the world. Kepler, too, used diagrams, but of a scientific character—optical constructions made up of rays of light, a sphere with light ema-nating from its center as straight lines, or an image of planets moving in ellipses around a sun displaced from the center of the universe.

Both agreed that there was an invisible realm of qualities and pow-ers, as well as a harmonics of nature. But while Fludd's world was one of astral powers and invisible spiritual illumination, Kepler's was of invisible magnetic forces, archetypal images, and hidden astrological meanings.

In his *Harmonices*, Kepler included a devastating critique of Fludd's book. Fludd immediately sprang to the defense. To start with, Kepler derided Fludd's extensive reliance on pictures; for what interested Kepler was mathematics. In reply, Fludd ranked him with "vulgar mathemati-cians" who concern themselves only with "quantitative shadows." Phi-losophers like himself, Fludd wrote, "comprehend the true core of natural bodies" rather than stripping nature bare with cold mathematics. Kepler replied, "In Fludd's method is the business of alchemists, hermetists and Paracelsians; mine is the task of the mathematician."

In his drawings Fludd represented the text of Genesis using images based on alchemy, astrology, and the Kabbalah, with light playing a central role. To him the mundane world was the mirror image of the invisible world of the Trinitarian God. He represented this as two equilateral trian-gles placed together and wrote beside the upper one: "That most divine

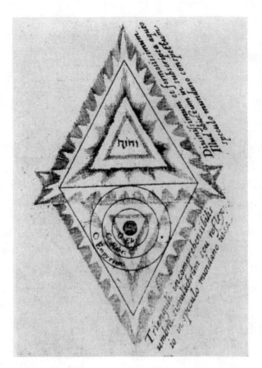

The divine and mundane triangles. (Fludd, Utriusque Cosmi Maioris scilicet et Minoris, Metaphysica, Physica atque technica Historices mundi *[1621].)*

and beautiful Object [God] seen in the murky mirror of the world drawn underneath." The upper triangle contains the four Hebrew characters יהוה—the tetragrammaton—for the ineffable name of God, YHVH, set within another perfect triangle. The triangle beneath it is the "reflection of the incomprehensible triangle seen in the mirror of the world," Fludd wrote.

To depict the creation Fludd used an image of interpenetrating triangles. One triangle ascends from the earth. It is dark at the base and becomes brighter as it moves toward heaven. The inverted triangle, meanwhile, has its apex on the earth. The former culminates in the perfect triangle, the symbol of God, while the latter emanates from it. They mirror each other precisely and thus represent the constant struggle of polar opposites: the triangle rising from the earth represents the dark principle, or *matter,* while the other is the light principle, or *form. Matter* and *form,* light and darkness are the polar principles of the universe. This is reminiscent of the Kabbalah where these opposites are called *antipathy* and *sympathy.*

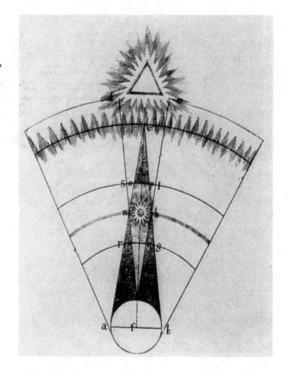

Interpenetration of material and formal pyramids. (Fludd, Utriusque Cosmi Maioris scilicet et Minoris, Metaphysica, Physica atque technica Historices mundi *[1621].)*

Fludd emphasizes that the world about us results from a struggle between dark and light by placing the sun at the intersection point of the two pyramids, where the opposing principles counterbalance each other. This also signals his belief that the unity of God Himself is symbolized in the mystery of the alchemical wedding in which opposites are fused together.

Placing the apex of the light triangle on the earth symbolizes the withdrawal of light and the appearance of matter. In his analysis of all this, Pauli was particularly interested in the Lurianic story of Creation, as revealed by the sixteenth-century mystic and kabbalist Isaac Luria, of whom Fludd was aware. Luria reported that his soul often traveled to divine realms to study the secrets of existence and claimed that he could not write his visions down because they gushed so rapidly from his mind. Others recorded them in what became known as the Lurianic Kabbalah. Some of his disciples asserted that his early death, at thirty-eight, was God's retribution on him for revealing forbidden knowledge.

Luria asked questions such as, Why everything? Why did creation occur? What is the meaning of everything? Fludd's inverted triangles contain his replies. Luria called *Tsimtsum*—the withdrawal of light and thus of God to create matter—one of the most important notions in kabbalistic thought. The problem is, if God is everywhere, how can there be a world? How can there be anything that is not God? To accomplish this separation God must have had to abandon a region within Himself to create a "kind of mystical primordial space from which He withdrew in order to return to it"—or so the kabbalistic scholar Gershom Scholem, a friend of Pauli's, wrote.

Once darkness, or Nothing, could be visualized, then the act of creation—Let there be light!—followed, or so Fludd believed. To express this he drew a dark square. In another image he drew rays of light emanating from a dark core and terminating at a circular periphery with darkness outside it—light, dark, and spirit, the Trinity. From this triad, according to Fludd, the four elements emerged and the struggle among them began. This cosmogony was the blueprint for all natural processes in that they were bases for all subsequent alchemical transformations among the four elements.

Thus the Pythagorean tetraktys emerged out of Fludd's version of how God created the cosmos. First comes the unity (one) culminating in darkness, followed by the duality of light and dark (two), then by the Trinity (three), culminating in the four elements and the four seasons, and all the other sets of fours that make up the world as we know it, Fludd argued. Pauli wrote appropriately, "His goal is the coniunctio of light and darkness: not the spiritualization of matter. . . . This is alchemy in the best sense."

Kepler versus Fludd

Kepler scoffed at Fludd's attempt to seek harmonies "from the interpenetration of his Pyramids which he privately carries around in his mind as a world drawn in pictures." Kepler conversely claimed to have found harmonies in the motions of the planets within a scheme based on mathematics, and that fit astronomical observations and measurements. Without mathematics, he wrote, "I am like a blind man." While Fludd claimed to take his lead from the "Ancients," Kepler followed "Nature herself."

All the same, Kepler's *Harmonices* was full of astrological, alchemical, Pythagorean, and mystical concepts. Even though Kepler had fulfilled Pythagoras's dream of explaining the universe through geometry and number, he was not satisfied. He was torn between the irresistible pull of his three laws, which postulated that the sun was *not* at the center of the universe, and the archetypal Trinitarian view of a spherical cosmos with the sun at its center. They did not mirror each other.

He fretted over the division between inert and live matter. Mathematics seemed to apply only to the former; but surely matter had a soul? He could not derive his laws with the mathematics available to him and they did not make much sense without the concept of there being something that tied the planets to the sun.

Fludd published the full text of the *Macrocosm* two years after Kepler's *Harmonices*. "Spurred on by the insolence of" Kepler, Fludd gave the usual Pythagorean reasons as to why the key number of the universe was four: its importance for geometry and music, its role in the "mystery of the seven days of creation: the sun was created on the fourth day." Four plus three, he pointed out (the quaternary plus the Trinity) adds up to the magic number seven.

He then referred to the four letters that made up the name of Yahweh—יהוה—the tetragrammaton. The double "He," he wrote, signified the progression from the Father to the Son.

To this he added the "hieroglyphic monad"—the four symbols representing the sun, the moon, the four elements, and fire. These are depicted as the crescent moon on the round sun, connected by the "quaternary of the cross, four lines being arranged so as to meet in the common point" to the symbol for fire. All of these, according to ancient beliefs and also the beliefs held at the time—such as Hermeticism, alchemy, the Kabbalah, and the Rosicrucians—are responsible for the cycle of transformations that produce our world.

Kepler looked at all this and realized he was wasting his time. He decided to cease communicating with Fludd, "I have moved mountains; it is astonishing how much smoke they expel," he wrote.

All coherence gone

Among Kepler's last projects was the completion of *Somnium, Sive Astro-nomia Lunaris—Dream or Astronomy of the Moon*, a science fiction story about a journey to the moon. In it he imagined what the universe would look like to someone standing on the moon. It was a bold notion that had been important to his discovery of his three laws.

The *Somnium* in its fragmented form sparked the curiosity of many readers, including the poet John Donne. Donne visited Kepler in 1619 as part of an English delegation dispatched by King James I to Germany. He was interested also in Kepler's book on new stars, *De Stella nova*. Donne was struck by the implications of the new astronomy: stars no longer immutable, the earth no longer at the center of the universe and, worst of all, the universe most likely of infinite extent, making Heaven far away while Hell was just beneath our feet. "'Tis all in peeces, all cohaerence gone [*sic*]," he wrote.

At this point Kepler and his family were living in an apartment in the wall surrounding the city of Linz, which was constantly under siege. They often had to admit soldiers to fire their guns through their windows. When the siege was lifted in 1626, Kepler finally left. He died in Regens-burg on November 15, 1630. The cemetery in which he was buried was obliterated in the Thirty Years War.

Fludd died seven years later in London. Like a lightning rod his ideas had attracted sharp controversies, most notably with Kepler. His will stated that all those at his funeral should return to the local pub and entertain themselves at his expense.

Three or four?

To Pauli, Kepler and Fludd were a study in opposites. At first he sided with Kepler but over time came to realize that Fludd's worldview included science, music, religion, and the mind. For Fludd four was "the eternal fountainhead of nature." For Kepler the perfect number was three. "I hit upon Kepler as trinitarian and Fludd as quaternarian—and with their polemic, I felt an inner conflict resonate within myself. I have certain features of both," Pauli wrote.

Like Kepler, Pauli brooded about his work, suffering over problems he couldn't solve, far removed from the world of ordinary people. In 1924, when he discovered the exclusion principle, perhaps like Kepler he felt that he had tapped into something that went beyond science. Moving from three to four quantum numbers was a momentous step. It meant a complete break with the iconic imagery of the Bohr atom as a miniscule solar system. It was a step into the unknown, into a world without any visual images. Perhaps Pauli had in mind one of Bohr's favorite sayings from the eighteenth-century German poet Friedrich Schiller:

> Only fullness leads to clarity
> And truth lies in the abyss.

In his day Kepler stopped short at the number three, basing this decision on the three-dimensionality of space, on the one hand, and the Holy Trinity on the other. The deep mysteries of alchemy with its emphasis on the number four overwhelmed him.

6

Pauli, Heisenberg, and the Great Quantum Breakthrough

.

EVERYONE AGREED with Pauli that there should be four not three quantum numbers. His exclusion principle had shown that no two electrons in an atom could have the same four quantum numbers. Beyond that his colleagues could see that the principle must have huge implications. But no one could yet see what they were.

By the beginning of 1925 it was clear that Bohr's theory of the atom as a miniature solar system no longer provided even an adequate basis for understanding the atom, let alone for the exclusion principle or the anomalous Zeeman effect. Bohr's theory by now was under attack from all sides.

The demise of Bohr's theory of the atom

Pauli, despite his best intentions, had been one of the key wielders of the knife. As he had discovered, the theory had failed to produce a realistic model of either a hydrogen-molecule ion or a helium atom. Then new data appeared showing that the hydrogen atom did not respond to being hit by light as if it were a tiny solar system. This model produced spectral lines for the struck light that did not agree with those found in the laboratory.

Bohr fought back with a variation of his theory in which the invisible orbits of the invisible electrons were replaced by invisible electrons on springs, each emitting light at the frequency of an observed spectral line. To emphasize that these invisible electrons were an intermediate kind of reality, he referred to them as "virtual oscillators." Pauli wanted nothing to do with them. He had had enough of bizarre models and was totally discouraged.

Heisenberg, who was twenty-four, thought otherwise. Throughout spring 1925 he pushed Bohr's theory of virtual oscillators to its limits. But it failed. It seemed that Bohr's theory barely worked even for the hydrogen atom and even then no one really understood why. Atomic physics lay in ruins.

Many physicists spoke of their despair. Pauli did not respond well to crises and was becoming more and more depressed. He joked bitterly that physics was all wrong and wrote to Kronig, "I wish I were a film comedian or something similar and had never heard of physics." He hoped, he added, that "Bohr will rescue us with a new idea."

Around this time, Pauli wrote to Bohr about Heisenberg, "I always feel strange with him. When I think about his ideas, they seem dreadful to me and inwardly I swear about them. For he is very unphilosophical, he pays no attention to expressing clearly the fundamental assumptions and their connection with existing theories. But when I talk to him he pleases me very much and I see that he has all sorts of new arguments. . . . I believe that some time in the future he will greatly advance science." Pauli was to be proved right.

Unlike Pauli, Heisenberg thrived in periods of chaos. Far from despairing, he would go all out to find a solution. He welcomed the stretch of the imagination required by Bohr's virtual oscillators. He used his immense experience in every aspect of atomic physics, together with his natural audaciousness, spurred on by Pauli's critical comments— among them that he should deal only with quantities that can be measured in the laboratory, such as the energy and momentum of electrons, and avoid abstract concepts such as orbits of electrons. "We must adjust our concepts to experience," was the approach Pauli suggested. Heisenberg worked day and night and came up with a whole new atomic physics that was to become known as quantum mechanics. Full of excitement, Pauli wrote that Heisenberg's work gave him "new hope and a renewed enjoyment in life."

"We must adjust our concepts to experience"

Like every highly creative scientist of his era, Pauli was a philosophical opportunist. He picked and chose from whatever philosophy had to offer to tackle the problem at hand. Scientists use philosophy when they ask the deepest of questions, such as What constitutes a scientific theory? What sort of physical objects should it consider and how should it treat them? What is physical reality?

At the beginning of the twentieth century these questions became crucial when scientists had to contend with objects—such as electrons and atoms—that they could not actually see. Classical ways of understanding the world suddenly seemed insufficient. An intellectual tidal wave—the avant-garde—swept across Europe.

Scientific concepts, ways of thinking, and ways of knowing were all being re-examined. Einstein did so when he discovered his special theory of relativity in 1905. This upheaval in thinking pervaded the world outside science too. In 1907 Pablo Picasso launched cubism with his "Les Demoiselles d'Avignon" and in 1910 Wassily Kandinsky unveiled abstract expressionism. In 1913 Igor Stravinsky ruptured all the conventions of classical ballet with his "Rite of Spring." The postwar 1920s produced the twelve-tone music of Arnold Schönberg, Bauhaus architecture, and James Joyce's extraordinary novels, which encompassed everything from relativity to cubism. Meanwhile Freud and Jung were investigating the unconscious.

Pauli first encountered this ferment of ideas through his godfather, the positivist Ernst Mach. As a boy he was spellbound by the scientific equipment in Mach's apartment. Its ultimate purpose, said Mach, was to eliminate unreliable thinking—to demonstrate that the only thing that was really out there was what you can experience with your senses. The rest was all metaphysics—quite literally beyond physics and not worth considering, mere illusion.

Atoms could not be experienced with the senses. Did that mean they were merely "metaphysical," in Mach's pejorative sense? Were they not part of the elaborate scientific theories which made predictions that could be proved in the laboratory? According to Einstein's theory of relativity—Pauli's first scientific love—time turned out to depend on the motion of a clock and our world was four-dimensional, not three as everyone had always thought. The message of relativity theory seemed to be that scientists should look beyond what was immediately perceptible by the senses. It was to Einstein's disappointment that he failed ever to convince Mach to accept relativity theory.

In the light of relativity theory Mach's view seemed too restrictive. A group of young philosophers with strong scientific backgrounds began to meet in the coffeehouses of Vienna to discuss how to correct this situation, how to bring positivism into line with relativity theory. They called themselves the Vienna Circle and came up with a sophisticated version of positivism that they dubbed "logical positivism." Then they renamed it "logical empiricism": the word "empiricism" refers to experimental data (empirical data). Logical empiricism emphasized the role of mathematics in that a theory required a consistent logical or mathematical structure. Mach, on the other hand, regarded mathematics as merely an economical way to summarize experimental data.

In the view of the Vienna Circle a scientific theory had to be built on empirical data with the help of mathematics and had to generate predictions that could be tested in the laboratory. Science was a two-way street, beginning with data and ending with predictions that could be verified by data in the laboratory. Logical empiricism also insisted that every concept in a scientific theory must be measurable. Distance could be measured with a ruler, time by clocks, and so on. Thus they claimed that Einstein's discovery of relativity theory was actually in accordance with positivism.

As for atoms, this was just a name for a list of experimental results. The rays emerging from cathode-ray tubes—primitive television tubes— were assumed to be a sort of light ray with an electric charge. Actually, every scientist knew that cathode rays were made up of electrons. Both

Mach and the logical empiricists declared that atoms were not real as they could not be seen or measured individually. But the logical empiricists were able to see a way around Mach's rejection of Einstein's theory of the relativity of time in that it emerged from a consistent mathematics and experiments had been done to illustrate it in the laboratory. Mach's philosophical heirs made the important point that the criterion "to observe something in the laboratory" had to be replaced by "to ascertain it or measure it in the laboratory."

Pauli was well read in philosophy and introduced himself to the then-doyen of the Vienna Circle, the German-born Moritz Schlick. Schlick was twice his age and an esteemed professor at the University of Vienna, where he had taken over Mach's position. Schlick was impressed with Pauli's philosophical acumen. Pauli did not let the fact that he was a mere postdoctoral student hinder him from giving Schlick his blunt assessment of positivism. He had no objection to it, he wrote in 1922, "But, of course, it is not the only [philosophical approach]."

Indeed it was not. With the rise of psychoanalysis scientists began to look into how they had come up with their discoveries. Einstein wrote, "There is no logical path to these laws; only intuition, resting on sympathetic understanding of experience, can reach them." In this way, he added, scientists could glimpse the "pre-established" harmony of the universe. Logical empiricists, however, saw this as aimless babble conjured up by scientists years after the fact.

In their view scientists constructed theories by moving logically—mathematically—from experimental data to a theory. They churned out equation after equation until they had solved the problem at hand. Einstein considered this wrongheaded. Scientists were unanimous in agreeing that their methods of research bore no resemblance to the proposals of positivists and logical empiricists. The key point for creative scientists such as Einstein was the delicate balance they had to maintain between the information obtained from experimental data and the laws of the theory as expressed in mathematics.

Pauli undoubtedly read Einstein's views as well as the famous polemic in the first decade of the twentieth century between Mach and the discoverer of quantum theory, Max Planck, whose opinions were similar to Einstein's. Planck accused Mach of degrading physics whereupon Mach simply withdrew in disgust: "I cut myself off from the physicist's mode of thinking."

Einstein believed, as did many scientists, in a world beyond perceptions in which electrons actually existed. Philosophers called this view "scientific realism." There were scores of hybrid philosophies besides scientific realism and positivism which asserted that in fact there was nothing "out there." Pauli counted himself a " 'heretic,' not bowing down to any God, authority or 'ism.' "

As a philosophical opportunist, Pauli saw that positivism offered a way out of the morass of 1925, when Bohr's theory of the atom had collapsed with nothing to replace it. He thus advised Heisenberg to drop the unmeasurable concept of electron orbits and focus instead on measurable concepts like energy and momentum. This meant dropping the reassuring visual image of the atom as a solar system. Pauli's belief was that once the "systems of concepts are settled," that is, once the new atomic theory had been worked out, then "will *visual imagery* be regained," as he wrote to Bohr. At Bohr's Institute, Heisenberg and Bohr shared all correspondence from Pauli and eagerly awaited it.

Quantum mechanics—the new atomic physics

Heisenberg's quantum mechanics identified individual electrons within atoms by the radiation they emitted while jumping between different stationary states, that is, the condition of an electron characterized by four quantum numbers as well as its momentum and energy, measurable as spectral lines. The transitions, or jumps, of the electrons maintained the flavor of the discontinuous quantum jumps in Bohr's theory of the atom. Discontinuities were a fact of life in the world of the atom, especially in a theory based on electrons as particles.

Pauli was convinced that Heisenberg's quantum mechanics would make it possible to solve problems that he had been unable to solve with the old Bohr theory. Late in 1925, he set out to calculate the stationary states for the simplest atom—hydrogen—using quantum mechanics. It involved juggling very complex mathematics but he came up with the answer with amazing speed.

Werner Heisenberg in 1925, when he discovered quantum mechanics.

Bohr applauded Pauli's "wonderful results." Heisenberg complained he was "a bit unhappy" that he had not solved the problem himself, but was full of admiration and surprise that Pauli had done it "so quickly."

Irked that Pauli had stolen a march on him, just a month later Heisenberg, along with Pascual Jordan, another brilliant young physicist, tried applying quantum mechanics to the problem that had driven everyone to despair—the anomalous Zeeman effect. Just as Kepler's ellipses had eliminated the cumbersome circles moving on circles, so Pauli's new concept of spin—part of Heisenberg's quantum mechanics—at a stroke swept away the concepts of massive inert cores with their two-valuedness and strange forces which had cluttered up Bohr's theory. The problem had finally been put to rest, and the solution also helped set Heisenberg's quantum mechanics on a firm basis. This time they had the theory right.

Physicists applauded these calculational breakthroughs. But no one—including Bohr, Heisenberg, and Pauli—really understood the theory itself, because the properties of atomic entities were so impossible to imagine. Not only was it unfamiliar and difficult to use, the mathematics of Heisenberg's quantum mechanics lacked any helpful visual image. Being a hybrid version of Bohr's virtual oscillators, it was like trying to visualize infinity. Its fundamental particles were also unvisualizable. But this was fine with Heisenberg, who felt the time was not ripe for a return to visual images which, in the past, had always turned out to be misleading.

Then the French physicist Louis de Broglie suggested that electrons might be waves—in other words, that material objects, such as ourselves, might be considered as waves. His inspiration was Einstein's discovery, made some two decades previously, that light—traditionally thought of as a wave—could also be a particle, dubbed a light quantum. Perhaps electrons as well as light might be both wave and particle at the same time—simply beyond imaginable.

In spring 1926, the flamboyant Erwin Schrödinger, at the University of Zürich, burst on the scene. At thirty-nine, Schrödinger was an outsider in age, temperament, and thought to the group of impetuous twenty-something quantum physicists who clustered around Bohr in Copenhagen, Sommerfeld in Munich, and Born in Göttingen.

Schrödinger had found the equation that converted de Broglie's vision of matter as waves into a coherent theory. His version of atomic physics, which he called wave mechanics, was based on treating light and electrons as waves. "My theory was inspired by L. de Broglie," he wrote. "No genetic relation whatever with Heisenberg is known to me. I knew of his theory, of course, but felt discouraged, not to say repelled, by the methods of the transcendental algebra, which appeared very difficult to me, and by the lack of visualizability."

Schrödinger's wave mechanics sprang from a preference for a mathematics that was more familiar and beautiful, as opposed to what he referred to as Heisenberg's ugly "transcendental algebra." The "Schrödinger equation" offered great advantages in calculations over Heisenberg's quantum mechanics, added to which it enabled the electron in an atom to be visualized as a wave surrounding the nucleus. It had taken Pauli twenty-odd pages to solve the hydrogen atom problem. Schrödinger did it in a page.

Schrödinger pointed out that the wave nature of matter promised a return to classical continuity. The passage of an electron between stationary states could be envisioned as a string passing continuously from one mode of oscillation to another.

One year earlier there had been no viable atomic theory. Now there were two: Heisenberg's quantum mechanics and Schrödinger's wave mechanics.

Heisenberg was furious about Schrödinger's work and even more so about its rave reviews from the physics community. "The more I reflect on the physical portion of Schrödinger's theory, the more disgusting I find it," he wrote to Pauli. "What Schrödinger writes on the visualizability of his theory I consider crap."

Heisenberg saw wave functions—that is, the solutions to Schrödinger's wave equation—as nothing more than a means to expedite calculations. To demonstrate this he applied them to the problem that had driven Born, Pauli, and himself to despair: to find a mathematical way to describe the properties of the helium atom. No one had been able to deduce stable orbits, or stationary states, for the two electrons in the helium atom using Bohr's theory of the atom. This being the case, they could not move on to deduce spectral lines for the helium atom because these resulted from its electrons dropping down from a higher to a lower orbit. Instead, the electrons' orbits remained unstable, meaning that an electron could be knocked out of the helium atom by the smallest of disturbances.

But in Heisenberg's quantum mechanics there were no orbits. The problem became one of deducing the atom's spectral lines from its stationary states expressed directly in terms of the electrons' energy and momentum in the atom. If the spectral lines turned out to be wrong, it would show that there were serious problems with the way quantum mechanics defined stationary states, that is, the energy levels of electrons in atoms. The spectral lines of the helium atom were particularly interesting to physicists because, as had been observed in the laboratory, they fell into two distinct groups. But why should this be the case?

Insight into the exclusion principle

The helium atom has two electrons. Using his quantum mechanics Heisenberg showed how the two sets of spectra arise. To elucidate his result and speed up his calculation of numerical values for the spectral lines, he used Schrödinger's wave functions—the solutions to the Schrödinger equation—for both the spins and positions of these two electrons. The total wave function is the result of multiplying these two wave functions together. But there are many possible ways of constructing the total wave function for these two electrons.

Heisenberg found that only one sort produced the two distinct groups of spectral lines characteristic of the helium atom. This particular wave function had a unique property. It changed its sign when the spins and positions of the electrons were swapped. It was antisymmetrical, which also meant that it went to zero if the electrons had the same spins or positions.*

What was nature's selection device for choosing these two sets of wave functions for the two spectra out of the several possible ones? Heisenberg was stumped. Something strange was going on here. Perhaps it related to Pauli's exclusion principle, according to which no two electrons could have the same spin and position. If they did then one of the two wave functions that make up the total wave function—either for their positions or for their spins—would have to become zero. Perhaps that was the way nature selected the wave function suitable for a particular system of electrons. Thus Heisenberg realized that the exclusion principle was related to the symmetry property of the wave functions for a collection of electrons, in this case two electrons. It was a step forward in exploring its implications beyond making sense of the periodic table of elements.

It was a typical Heisenberg project. He chose a fundamental problem—in this case to understand the spectrum of the helium atom—and then let his intuition lead him into new terrain: the symmetry property of wave functions whether they are symmetric or antisymmetric. Thus he realized how essential the exclusion principle was for quantum mechanics: without it quantum mechanics could not be complete.

There was also the problem that had been Pauli's original bête noir

---

*The term *symmetry* has a very specific meaning in physics. When an equation remains the same even when its mathematical symbols are altered it is said to possess symmetry. The equation for a sphere remains the same if each point on the sphere is reversed; a sphere is still a sphere even when looked at in a mirror.

from his PhD thesis, in which he showed that Bohr's theory of the atom failed to produce a stable hydrogen-molecule ion, H_2^+, even though it existed in nature. This problem vexed Born and Heisenberg as well. Pauli wrote to his friend Wentzel, "In Copenhagen sits a gentleman who is calculating H_2^+ according to Schrödinger." The "gentleman" was the Danish physicist Øvind Burrau who, as Pauli pointed out, started directly with Schrödinger's wave mechanics as opposed to starting from the quantum mechanics as Heisenberg had done and used Schrödinger's wave mechanics only for calculations. As a result he was able to solve the problem simply. Heisenberg wrote to Pauli that, in his opinion, Burrau had straightened out the situation and mentioned the symmetry properties of the wave functions that Burrau had deduced. Perhaps Heisenberg had hoped to find a solution starting from his quantum mechanics. But these once-key problems had become mere calculations now that the correct atomic physics had been worked out.

Although problem after problem that had resisted solution using the old Bohr theory was now being solved by atomic physics, the meaning of the theories used—Heisenberg's quantum mechanics and Schrödinger's wave mechanics—was still not understood. And the tension between the two factions was growing.

To the Schrödinger faction Burrau's successful result, as well as Heisenberg's for the helium atom (despite his assertion that he had used Schrödinger's theory merely to facilitate calculations) was proof that Schrödinger's theory offered a solution to every problem of atomic structure, whereas Heisenberg's was daunting to use and ugly. This of course greatly pleased Schrödinger.

Heisenberg's uncertainty principle

In fall 1926, Bohr summoned Heisenberg to his institute in Copenhagen to hammer out a resolution to the dispute with Schrödinger. They struggled for days over numerous cups of tea and bottles of Carlsberg beer. Heisenberg wrote to Pauli: "What the words 'wave' or 'particle' mean we know not any more; [we are in a] state of almost complete despair."

The crux of the problem was this: how could ordinary language, with its visual connotations, be used to describe a realm of nature that defied the imagination?

While Bohr and Heisenberg were deliberating in Copenhagen, Pauli had an idea. He immediately wrote it up and mailed it to Heisenberg. It was based on an insight Born had had, looking into Schrödinger's wave equation.

Born had suggested that the wave function was a wave of probability for an electron moving between stationary states. Pauli pushed the idea further. He realized that the wave function gave the probability of an electron being detected in a certain region of space. In his usual way, he didn't bother to write a paper to publish this idea and in the end Born took the credit.

But as he was working out the mathematics for Heisenberg, he came up with another extraordinary discovery: if he could determine a particle's position accurately, he could not determine its momentum with the same accuracy. Pauli was puzzled as to why this should be so. Why couldn't he determine both with the same degree of accuracy? Heisenberg was struck by the insight. He was, he wrote to Pauli, "more and more inspired by the content of your last letter every time I reflect on it."

By February 1927 Bohr and Heisenberg had hit an impasse in their discussions on the deep meaning of the quantum theory, which seemed to be riddled with ambiguities. Bohr took a skiing break. Left to his own devices, Heisenberg set to work. The result was a paper that he called "On the Intuitive Content of the Quantum-Theoretical Kinematics and Mechanics." Hidden behind this daunting title was one of the most earth-shattering discoveries of modern physics: the uncertainty principle.

Heisenberg realized that the supposed ambiguities of the quantum theory were essentially a problem of language. The issue was how to define words such as "position" and "velocity" in the ambiguous realm of the atom, a world in which "things" can be both wave and particle at the same time. He used the term "intuitive" in the title of his paper, for his goal was to redefine the word in the world of the atom.

Certain concepts in quantum physics, Heisenberg claimed, such as "position" and "momentum" (mass times velocity), were "derivable neither from our laws of thought nor from experiment." Instead we would have to look into the peculiar mathematics of quantum mechanics, which should have alerted us all along that in the realm of the atom we would have to apply such words with great care.

In his paper Heisenberg made the amazing assertion that the more accurately we measure an electron's momentum in a certain experiment, the less accurately we can measure its position in that experiment. This quickly became known as the uncertainty principle. It was earth-shattering in that it questioned our understanding of the inherent nature of the physical world as completely as Einstein's relativity theory.

In the classical physics of Newton we can measure the position and momentum of an object with the same degree of accuracy by observing how it moves. Using a telescope and a clock we can measure both the location of a falling stone and how fast it is moving with an accuracy limited only by the width of the telescope's crosshairs and the clock's mechanism. If we make these errors as small as possible we can deduce very precisely the stone's position and momentum. In principle, the product of the errors in measurements of position and momentum can both be zero. In quantum mechanics this is not possible.

Heisenberg wrote all this down in a detailed fourteen-page letter to Pauli. He asked for "severe criticism"; after all, it was Pauli who had given him the key idea. Pauli was elated. "It becomes day in the quantum theory," he declared.

Bohr's complementarity and beyond

Bohr, however, was furious. He refused to let Heisenberg publish his paper on the subject, saying that Heisenberg had not provided any firm foundation for his argument. Furthermore, Heisenberg had based the argument entirely on the assumption that light and electrons behaved like particles.

Bohr insisted that electrons and light be understood as both wave and particle, even though this could not be imagined. One could visualize electrons and light as either a wave or a particle so long as one remembered the restrictions required by quantum mechanics, among them

Heisenberg's uncertainty principle understood within the larger context of waves and particles.

This meant that electrons in experiments could exhibit one aspect or the other, but not both at once. If one experimented on an electron as if it were a wave, that was what it would be for the duration of the experiment, and similarly if one treated it as a particle. Bohr called this "complementarity."

Bohr was convinced that complementarity was relevant not only to physics but also to psychology and to life itself. Its basic idea, he wrote, "bears a deep-going analogy to the general difficulty in the formation of human ideas, inherent in the distinction between subject and object." As in the Chinese concept of yin and yang, complementary pairs of concepts defined reality. There is nothing paradoxical about an electron having the characteristics of both a wave and a particle until an experiment is performed on it. It dawned on Bohr that in the weird quantum world there need not be only yes and no, an electron need not actually be either particle or wave. There could be in-betweens as well as ambiguities. An electron's wave and particle aspects complement each other, and their totality makes up what the electron is. Thus the electron is made up of complementary pairs—wave and particle, and position and momentum. Similarly it is the tension between complementary pairs—love and hate, life and death, light and darkness—that shapes our everyday existence.

Bohr sent the manuscript of his article on complementarity to Pauli for corrections and critical remarks. Pauli replied immediately. Apart from certain comments on details, he entirely agreed with Bohr's thesis.

Only the more philosophically inclined scientists took complementarity seriously. Pauli was one. He began to look to complementarity as another way to study consciousness as in the various ways of "knowing" practiced in the East and West. He was growing more and more interested in the conscious and the unconscious, the rational and the irrational, and in how physics could be used to understand these complementarities. He was beginning to suspect that this was to be his life's work. The only problem was how to approach it.

Paul Dirac and quantum electrodynamics

The previous autumn the eccentric twenty-five-year-old English physicist Paul Dirac had visited the Bohr Institute. Dirac had already made important contributions to atomic physics and was eager to rub shoulders with other physicists of his generation whose papers he had studied in detail, Heisenberg and Pauli among them.

Dirac had been privy to the intense conversations between Bohr and Heisenberg on the issue of whether light and matter could be both wave and particle. In 1927 he was able to provide the vital clarification through a mathematical method he had developed for moving between the two and thus brought about "complete harmony between the wave and light quantum descriptions." Dirac's mathematical method ultimately concerned the way in which electrons and light interact. It formed the basis for a whole new subject, which scientists dubbed quantum electrodynamics. Pauli and Heisenberg worked enthusiastically to develop this new field.

Dirac's equation

The following year Dirac came up with a crucial equation—the Dirac equation. It described how electrons interacted with light and also agreed with relativity theory. The equations in Heisenberg's and Schrödinger's theories did not agree with relativity, and while the spin of the electron had to be inserted into these theories, it popped right out of Dirac's, thus underscoring the relationship of spin to relativity.

But Pauli and his colleagues were dissatisfied. Among other problems, Dirac's equation for the electron predicted that objects with negative energy should actually exist. Physicists believed particles of negative energy to be like negative time: they simply could not exist. Heisenberg commented to Pauli that Dirac's equation was the "saddest chapter in modern physics." Furthermore, it did nothing to elucidate Pauli's exclusion principle.

Pauli's anti-Dirac equation

In 1934 Pauli set out to find an equation to supplant Dirac's. His colleague in this was his assistant Victor Weisskopf. Weisskopf had been a student of Born's and Bohr's. Like Pauli, he was Viennese. The two shared a deep appreciation for literature and music, in particular Mozart's operas. Weisskopf was also a concert-level pianist. Well over six feet tall, athletically built, and with a cultured air, he stands out in group photographs.

Viki, as Weisskopf became known to his friends, loved to tell the story of his first meeting with Pauli in the fall of 1933:

> The first time I came to see him, I knocked at the door—no answer. He was in a very bad mood at that time, the whole period was a difficult one for him for personal reasons. When he didn't answer, after a few minutes I opened the door. 'You are Weisskopf; yes, you will be my assistant. I will tell you that I wanted to take [Hans] Bethe but he works on solid state [physics]. I don't like this kind of physics although I started it.' He gave me some problem . . . and after a few weeks I showed him what I had done; he was very dissatisfied with it and he said, 'I should have taken Bethe.'

This was Pauli's idea of humor, but it also shows what a difficult, sharp-tongued man he could be. Many people could not handle his sardonic sense of humor and this led, no doubt, to some physicists thinking twice before taking on the job of his assistant. But despite the inauspicious start Pauli's collaboration with Weisskopf was to be both fruitful and memorable.

The two came up with an equation that had many of the same properties as Dirac's and agreed with relativity theory. But while Dirac's was for any particle with half a unit of spin, theirs was for particles with no spin. None had been detected at the time. However, when they included spin in their equation it no longer agreed with relativity.

But why? As he was fiddling about with his equations Pauli realized something entirely new: that particles with no spin differ fundamentally from particles with half a unit of spin. Particles with half a unit of spin—1/2, 3/2 and so on (known as Fermions after the Italian physicist Enrico Fermi)—obeyed Fermi-Dirac statistics (discovered by Fermi and Dirac in 1926), meaning that the overall wave function (that is, the solution of the Schrödinger equation) for a collection of Fermions exhibited antisymmetry.[*]

The only known such particles at the time were the electron, neutron, neutrino, and proton. Particles with whole-number spin—zero, one, and so on (called Bosons after the Indian physicist Satyendra Nath Bose)—obey Bose-Einstein statistics (discovered by Bose and Einstein in 1924), meaning that they have an overall wave function that remained the same when their positions and spins are exchanged; it was symmetrical. In other words, the two sets of particles have different symmetry properties. From this Pauli deduced that the exclusion principle applies to particles with half a unit of spin and not to particles with a whole unit of spin.

There was no obvious reason for this. The only conclusion was that nature had spoken. Something more than mathematics was involved in wave equations. Physicists now began investigating the properties of wave equations for particles of any sort of spin. The difficult mathematics involved was very much to Pauli's taste.

Six years later, in 1940, he summarized and extended this work. Instead of using any particular wave equation such as Dirac's, his own, or any other, he used the mathematical properties of wave functions to explore how they behaved when relativity theory, spin, statistics, and the exclusion principle were applied to them. He cut through all the mathematics to deduce a conclusion that was highly significant. The exclusion principle, he discovered, cannot be applied to any theory that includes relativity and applies to particles with whole-number spin; but it is essential to theories dealing with particles with half-units of spin.

The "connection between spin and statistics is one of the most important applications of special relativity theory," he wrote. He had been trying for a long time to find a connection between spin, statistics, and relativity and at last he had done it. Compact yet rigorous in its

---

[*]In quantum physics "statistics" can refer to whether the wave function of a collection of particles is symmetrical or antisymmetrical when their spins and positions are swapped, which, as we saw, was first noted by Heisenberg.

mathematical presentation, the paper he wrote on the subject was Pauli at his best. Many physicists regard it as his most brilliant.

Finally, some sixteen years after Pauli had first come up with the exclusion principle and with the concept of spin—and had first realized that there were four, not three quantum numbers—he had managed to discover some of the key implications of his first great discovery. From the start everyone had realized that the exclusion principle explained the periodic table of elements. Now it was known that it could be used as a tool to explore the behavior of every particle with half a unit of spin and that it had no connection with any other sort of particle.

7

Mephistopheles

God's whip

PAULI never failed to give full rein to his sardonic humor and caustic wit. He ruthlessly criticized people who he thought did not think clearly. In 1926 he was at a lecture given by the Dutch physicist Paul Ehrenfest. Ehrenfest, a senior figure whose circle included Einstein and Bohr, was famous for his profound understanding of physics and his excellence at explaining difficult concepts. A charismatic teacher, he had motivated a succession of brilliant young students, but he seemed never to make great discoveries to match those of his famous colleagues, and the comparison tortured him. To him Pauli was one of those young "smartalecks. . . . Always so clever they were! And nobody understood anything."

After his lecture Pauli offered a string of critical comments. Finally Ehrenfest retorted, "I like your publications better than I like you." "Strange. My feeling about you is just the opposite," Pauli countered. The two later became good friends and Ehrenfest came to admire Pauli's critical acumen. "There is in *rebus physicus* only ONE God's whip (Thank God!!!)," Ehrenfest wrote to Pauli a couple of years later. Pauli was delighted with the title Ehrenfest had bestowed on him and thereafter often signed himself "God's whip."

Other acid quips have since become part of Pauli's lore. When students or colleagues tried out a new theory on him, he would sometimes shout, "Why, that's not even wrong," meaning that it was so far from correct it wasn't even possible to judge it by the normal standards of right and wrong. Other favorite one-liners of his included, "You're no more interesting drunk than sober," and "So young and already so unknown." Colleagues remember that Pauli liked to dream up a cutting remark, keep it in mind, and then, at the appropriate time, use it.

Pauli even criticized Bohr, who sometimes took offense. To Einstein, "I will not provoke you to contradict me, in order not to delay the natural death of [your present] theory," he wrote, about one of Einstein's attempts at a unified theory of gravitation and electromagnetism. Einstein, conversely, was quick to compliment the younger man on pointing out an error in another attempt at a unified theory. "So you were right, you rascal."

The one person Pauli never criticized was his mentor Sommerfeld. Pauli once wrote of the "awe you instilled in me . . . not even accorded Bohr." In his presence Pauli was a completely different person. Whenever Sommerfeld visited the Swiss Polytechnic Institute, later referred to as the ETH, in Zürich, where Pauli was working, Weisskopf recalled Pauli would always be saying, "Yes, Herr Geheimrat [Honored Teacher], yes, this is most interesting, but perhaps you would prefer a slightly different formulation, may I formulate it this way."

At base, Pauli's razor-edged criticism arose from his dislike of shoddy thinking. As a child he had been raised in a house in which everything from science to politics was pulled apart and criticized. "Obedience to authority was not sung to me in the cradle," he once wrote. But he simply said what was on his mind. He did not mean his criticisms to be taken personally and he was hardest of all on himself. In a letter to Pauli asking his advice, Heisenberg referred to him as the "master of criticism," and later recalled that "I have never published a work without having Pauli read it first." He was often called the "conscience of physics."

Fresh start

In professional terms, 1927 was a year of immense achievement for Pauli. He had been instrumental in helping Bohr and Heisenberg straighten out quantum theory and was engaged in working with Heisenberg in developing the field that Paul Dirac had initiated, quantum electrodynamics. Then came a shock.

Pauli's father had always been a womanizer. It was a situation that his mother, the brilliant intellectual journalist, Bertha, had had no choice but to resign herself to. Late that autumn, Wolfgang Sr. finally left her. He had fallen in love with a woman Pauli's age, a sculptor named Maria Rottler whom Pauli referred to caustically as his "wicked stepmother." No doubt the younger woman with her artistic ambitions offered Pauli's father a new lease on life, but to Pauli the desertion of his mother was an unforgivable act of treachery and betrayal.

It was more than Pauli's poor mother could bear. Not long after her husband left, on November 15, 1927, she took poison and died. Clearly for Pauli it was an unbearable trauma. He closed up and said not a word about it to his friends or colleagues. The extent to which it affected his mental well-being only became clear much later, when he began analysis with Jung.

The same month that Pauli received news of his mother's death, he also received another, more welcome, communication: the offer of a position at the prestigious ETH. The new job would mean leaving Hamburg for Zürich.

It happened that both the ETH and the University of Zürich were about to lose their most formidable theoretical physicists. Peter Debye, at the ETH, had accepted a position at the University of Leipzig while Schrödinger, at the university, had agreed to succeed Max Planck at the University of Berlin. Debye had been a student of Sommerfeld's. His main interest was investigating the structure of molecules by studying how they behaved when struck by x-rays, work that was later to earn him a Nobel Prize. The ETH decided to offer the vacant post to one of the two rising stars of theoretical physics—Pauli and Heisenberg. First Heisenberg was offered a position at the ETH, one of several offers he had that year. In the end, however, Debye lured him to Leipzig where he became a professor.

Next the ETH turned to Pauli. The eccentricities of his teaching style were well known. Valentine Telegdi remembered them as "pedagogically maladroit, but full of gems of wisdom which one had to find (and polish) oneself." Only the most brilliant students could understand anything—though for them it was a lesson not only in physics, but also in how to think critically about the subject. As Markus Fierz, later Pauli's assistant and then close friend, described it:

> His presentation was more like a soliloquy than a lecture. He spoke with an unclear twangy voice, and he wrote with small untidy letters on the blackboard. Sometimes he would lose the thread or, doubting the correctness of a derivation or a statement, shake his head or gaze noddingly into the air. He then continued, mumbling unintelligible words or saying "yes, yes, yes," though nobody knew what had disturbed him in the first place. This seemed to me extremely mysterious, and it contributed to intensifying the demonic aura surrounding this queer man.

The president of the ETH went personally to Hamburg to assess Pauli's performance. He decided that Pauli was young enough to improve and immediately offered him a position, starting on April 1, 1928, with a contract for ten years.

Zürich

"In April 1928 I arrived in Zürich as a new professor, dressed like a tourist with a rucksack on my back." Pauli went straight to his office in the imposing physics building at number 5 Gloriastrasse, a broad street off the main avenue, Rämistrasse, on which sits the principal building of the ETH. There he met his new colleague Paul Scherrer, who had had the office spruced up for the new professor's first day.

At twenty-eight, Scherrer, a debonair experimental physicist, was Pauli's exact contemporary. He had carried out important research on the x-ray analysis of crystals with Debye before joining the ETH in 1920 and, by 1927, had been promoted to head of the physics department. Besides his brilliance in research Scherrer was a charismatic lecturer. Twice a week he gave lectures to explain difficult concepts in physics to both scientists and laypeople, backed up by an array of spectacular demonstrations. The auditorium was always packed for what became known as the "Scherrer circus." On one occasion Scherrer tried out one of his simplified explanations on Pauli, who replied in his caustic way, "*Ja*, simple it is all right, but also wrong."

Pauli rented a flat on Schmelzbergstrasse, 34, a steep, narrow, tree-lined road, in a pleasant three-story house set back amidst trees, a few minutes walk from the physics department.

He immediately hired Kronig, who had been the first to realize that Pauli's fourth quantum number was the spin of the electron, as his assistant. "Every time I say something, contradict me with detailed arguments," Pauli told him. Next Pauli convinced his closest friend, the physicist Gregor Wentzel, to take the position vacated by Schrödinger at the University of Zürich. Like Pauli, Wentzel had been a student of Sommerfeld's in Munich. When Pauli was a student Wentzel had been Sommerfeld's assistant and joined in their café conversations. Two years Pauli's senior, he had already made important contributions to the new atomic physics. He had an easy-going manner, enhanced by his ever-present cigar and readiness for a good time. In letters Pauli addressed him as "Dear Gregor" and signed himself "Wolfgang"—in those days an extraordinary degree of informality.

Pauli had barely settled into the ETH when he was back at work with Heisenberg. But as the two resumed their research on quantum electrodynamics, they encountered such difficulties that Pauli began to lose heart. He lapsed into the state of mind he had had when previous problems proved intransigent—like in 1924, when his frustration over the anomalous Zeeman effect drove him to seek solace in the bars and whorehouses of the Sankt Pauli district, and again in 1925, when the Bohr theory of the atom collapsed and he daydreamed about giving it all up and becoming a film comedian. This time he played with the idea of dropping out to write a utopian novel. In reality he just needed a rest from physics.

He wrote to Bohr that he was having trouble concentrating. He wished he could say that he had no time for research or that he was tired, he added, but neither was true. "I am only stupid and lazy. I think that somebody ought to give me a daily thrashing! But since unfortunately there is no one around to do it, I must seek other means to reinvigorate my interest in physics." On the weekends he often went to Leipzig, where regional meetings of the German Physical Society were held. Heisenberg, too, was there and he and Pauli had endless discussions about issues holding up their progress in quantum electrodynamics.

One problem was that Pauli was distracted by the entertainment Zürich offered. He also had just the right mix of colleagues to share it with—Kronig and Scherrer. On warm Sundays the three friends swam at the Strandbad, a beach on Lake Zürich, a ten-minute drive from the city. To cap off the afternoon they would then go back to the city and walk along the elegant Bahnhofstrasse to Paradeplatz where they went to Sprüngli's café for ice cream and coffee. The rotund Pauli assigned Kronig the task of monitoring his ice cream intake.

In the evenings, after work, they walked down Rämistrasse to Bellevue Square, at the intersection of the Limmatquai, which runs alongside the river Limmat, and the Utoquai, to their favorite restaurants. The most elegant was the Kronenhalle, usually reserved for after-concert dinners. The dining room still retains its old-world flavor with its high ceiling, tables covered with white tablecloths, and waiters gliding around in black trousers, white shirts, black ties, and long white aprons tied at the back. Across the street and around the corner on Limmatquai is the Café Odéon. A masterpiece of art deco design, it was the meeting place for artists, poets, and intellectuals of every persuasion, even the occasional anarchist such as Lenin, who brooded and plotted there until the Allies returned him to Russia in a sealed train in 1917, as if he were a plague bacillus.

Across the Limmatquai is the Café Terrasse where the physics department took its colloquium speakers for dinner. It had a more intellectual atmosphere than the Kronenhalle and the discussion often continued there. In those days the now-enclosed dining room was an outdoor garden. Another spot Pauli and his friends frequented with some regularity was the Bauschänzl beer garden, just across the river Limmat from the

Café Terrasse, today a rather garish and pricey restaurant. Other bars and cafés where they spent time, such as the Café Voltaire, the headquarters for dadaism, are gone.

One warm summer's night after a day of swimming in the lake, ice cream at Sprüngli's, and dinner at Café Terrasse, the three friends sat in the Café Odéon, penning a postcard to Pauli's friend and successor in Hamburg, Pascual Jordan, who had worked with Heisenberg to crack the puzzle of the anomalous Zeeman effect. In the three years between his appointment as Born's assistant at Göttingen in 1924 and his arrival at Hamburg, Jordan had carried out ground-breaking work on the new quantum mechanics. Pauli addressed his postcard to "PQ – QP Jordan," a joking reference to an important equation Jordan had helped establish. "We are about to study the Zürich night life and try to improve it following the new method due to Pauli: by comparison," the three wrote exuberantly. "Many Greetings, Kronig," wrote Kronig. "This method, however, may also be used to worsen matters!—Greetings, Pauli," added Pauli. "I, too, have heard so many bad things about you that I would like to meet you. Scherrer."

Shortly afterward Kronig left for a position at Utrecht in the Netherlands and was replaced by Felix Bloch who had been Heisenberg's first PhD student. Tall, handsome, and urbane, Bloch fit in well with the crowd. Other soon-to-be-famous young physicists passed through the ETH, among them Rudolf Peierls, a student of Sommerfeld's and Heisenberg's, who succeeded Bloch as Pauli's assistant. A young man with immaculately parted hair and round-rimmed glasses, he had to suffer the sting of Pauli's famous biting remarks. One was particularly memorable: "[Peierls] talks so fast that by the time you understand what he is saying, he is already asserting the opposite."

J. Robert Oppenheimer worked with Pauli during the first half of 1929. Pauli wrote of Oppenheimer that he treated "physics as an avocation and psychoanalysis as a vocation." Perhaps Pauli sensed in Oppenheimer's complex and tortured personality a reflection of himself.

Motivated by his new colleagues, brilliant assistants, and a coterie of exceptional postdoctoral students, Pauli re-entered the stalled collaboration with Heisenberg. By September 1929 they had completed their opus, which they published in two parts. Its eighty-four pages firmly established quantum electrodynamics as a field of research and combined Heisenberg's intuitive approach and Pauli's penchant for rigorous calculations to dazzling effect. Covered with lengthy equations, the papers contained a wealth of new mathematical techniques that have since become part of every physicist's repertoire for exploring how elementary particles interact. "Not for the curious," quipped Pauli.

As well as its scientific importance, Zürich was a center of German culture and Pauli became a habitué of its intellectual salons. There he met, among others, the philosopher Bernard von Brentano; the writers James Joyce and Thomas Mann; Waldimir Rosenbaum, a wealthy political activist and lawyer; and the artists Max Ernst and Hermann Haller, the son of Einstein's stern but beloved former boss at the Swiss Federal Patent Office, Kurt Haller. Haller made a bust of Pauli that stands in the *La Salle Pauli* at CERN. He looks as if he is in deep contemplation, pondering weighty problems. "The sculptor Haller in Zürich has made a bust which makes me look rather introspective—i.e., Buddha-like," Pauli wrote to Kronig rather proudly.

Pauli in love

Amid this whirl of nightlife and salons Pauli was also building up the prestige of his department and conducting intense physics research. But no matter how much he filled his days and nights, he still anguished over his mother's death. It was only in 1929 that he hinted at his feelings, signing the papers on quantum electrodynamics that he had written with Heisenberg simply "Wolfgang Pauli," omitting the suffix "Jr." He no longer cared whether he was confused with his famous father whom he now loathed.

Then in May 1929 he made a rather strange decision that may or may not have had something to do with his mother's death: he left the Catholic church and unofficially adopted his father's original religion, Judaism. Perhaps it was his response to the harsh judgment of Catholicism, which condemns suicide as a mortal sin, that led him to take this step. He later described himself as a "Jew from the waist up." This may seem odd, in light of his ill feelings toward his father. But he had little regard for either Catholicism or Protestantism, to which his parents had converted in 1911, and he was well aware of how Jewish he looked, with his swarthy complexion, dark wavy hair, and dark eyes.

For all its pleasures Zürich had no Sankt Pauli red-light district where Pauli could seek consolation for his sorrows. He took to making frequent trips back to Hamburg and Berlin. Then, in December 1929, he suddenly announced, to the amazement of his colleagues, that he was going to marry—and, not only that, but that his intended was a cabaret dancer. He had always dismissed marriage as a bourgeois institution.

Perhaps the Zürich academics were surprised that no one else's wife was involved. In those circles extramarital affairs were common and most marriages quite open. Schrödinger, for example, had had a legendary number of liaisons. He usually traveled with both his wife and current girlfriend. Schrödinger's wife, meanwhile, was infatuated with the elegant mathematician Hermann Weyl, whose own wife was having an affair with Scherrer.

Pauli had fallen in love. He had met Käthe Deppner some years earlier, during one of his jaunts into the demi-monde of Berlin. Born in Leipzig, she was six years his junior and had trained at the famous Max Reinhardt School for Film and Theater in Berlin. Reinhardt had brought high-class theatre to Berlin and established a film industry that served as a model for Hollywood. Pauli's sister, Hertha, was with the Max Reinhardt theater and he later revealed that it was when he went to visit her that he had met Käthe for the first time. The two women were friends.

At first Pauli bowled Käthe over by boasting what an eminent physicist he was. Shortly afterward he ran into her again at a party in Zürich given by his friend Adolf Guggenbühl, a wealthy publisher and one of the founders of the magazine *Schweizer Spiegel*. Käthe was performing with a dance school founded by Trudi Schoop, an old girlfriend of Scherrer's. Zürich was a small world.

Why anyone falls in love is difficult to work out, especially when no hints are left behind, as was the case with Pauli and Käthe. Käthe was an attractive woman, with a round face and curly hair and a keen sense of fashion. No doubt as a chorus girl she had all the allure of the demi-monde and knew how to attract men. In a photograph of the two on a walk in the mountains, Pauli looks ecstatically happy. Perhaps he sensed a

Käthe Deppner and Pauli in the
countryside, 1929.

mutual longing and on impulse proposed marriage. It is harder to fathom why she accepted.

For even before the marriage had taken place, it was clear it was going to be a disaster. Käthe made it known that she had already fallen in love with someone else, but for some reason she decided to go through with the marriage to Pauli. Maybe she hoped it would work out in the end. After the ceremony the couple moved into a flat in Hadlaubstrasse, 41, a three-story house with large French windows on a pleasant tree-lined street.

Letters of congratulation poured in. But only two months later, Pauli was already confessing to a friend that he was married only in a very "loose way." He was clearly unhappy. "He used to walk around like a caged lion in our apartment," Käthe recalled, "formulating his answers [to letters] in the most biting and witty manner possible. This gave him great satisfaction." Pauli's profession demanded that he sit at a desk alone for hours on end. It must have been stifling for someone like her.

In November 1930, less than a year after they had married, the two divorced. Käthe had walked out on Pauli. What he most resented, he always said wryly, was that she had left him for a mere chemist. "If it had been a bullfighter—with someone like that I could not have competed— but with an average chemist?" he would complain. "In spite of his theories he is, like other mortals, at times vehemently plagued by jealousy," a colleague observed. Soon afterward Käthe married her chemist, Paul Goldfinger.

The lonely neutrino—Pauli's second breakthrough

Through all this tumult, Pauli's scientific creativity never flagged. The problem absorbing him at the time was, What happens when the nuclei of certain atoms give off excess energy by emitting an electron? Precise measurements of this process—known as beta-decay—showed, inexplicably, that the energy contained in the nucleus before the electron was emitted was greater than the combined energies of the nucleus afterward plus the discharged electron.

Somewhere, somehow, some energy had been lost. Could it be that beta-decay violated the law of conservation of energy, which holds that these energies must be equal? The law of conservation of energy was a mainstay of physics and engineering, and theories that violated it invariably turned out to be wrong.

Yet, in 1929, Niels Bohr had gone so far as to suggest that this fundamental law might not hold precisely for processes inside the nucleus, but only on average. "We must still be prepared for new surprises" in the atomic world, he wrote. Pauli and most other physicists strongly disagreed.

Pauli jokingly suggested to Bohr, "What if someone owed you a great deal of money and offered to pay it back in instalments, but each time the agreed instalment was not met? Would you consider this to be a statistical error or that something was missing?" In other words, Bohr was trying to make out that the missing energy was a matter of statistics and so was only missing on average.

In the papers on quantum electrodynamics that Pauli wrote with

Heisenberg in 1929, he had demonstrated that the law of conservation of energy was built into its equations. Now he pondered long and hard over the loss of energy in beta-decay. Then he came up with an audacious suggestion. Perhaps something really *was* missing. Could it be that the nucleus undergoing beta-decay emitted another as-yet-undetected particle, along with the electron, that would balance the books?

From the mathematics of beta-decay Pauli inferred that this particle's mass must be not more than the electron's, its spin must be one-half and it must have no electric charge. He was later to write of it as "That foolish child of the crisis of my life (1930–1)—which further behaved foolishly."

He announced his hypothesis in a letter to the audience at the radioactivity session of a physics meeting in Tübingen, Germany, in December 1930, a mere month after his divorce. Even as he pursued his scientific research he could not ignore the fact that his personal life was crumbling around him. The letter—beginning "Dear Radioactive Ladies and Gentlemen"—was to be read in his absence. Pauli had opted instead to attend Zürich's major social event of the winter, a ball at the splendid Baur au Lac.

In 1930 it was unheard of to suggest a new particle. No one had ever before dared do so. Were the electron, proton (the nucleus of the hydrogen atom), and light quantum (a particle of light) not enough? At first the scientific community was shocked. But it did not take long before everyone acknowledged that Pauli was almost certainly right.

A few years later the Italian physicist Enrico Fermi dubbed Pauli's new particle the neutrino. The neutrino was the centerpiece of Fermi's 1934 theory of beta-decay. One of the implications was how weakly neutrinos interacted with matter. The neutrino was a loner, it passed through the earth as if it were not even there and could whiz through space alone, not interacting with anything for three trillion miles. Yet neutrinos also constituted an essential part of the universe, required by basic laws of nature.

Soon after Pauli's hypothesis of the neutrino, experimental evidence on beta-decay suggested that if the neutrino existed, its mass would have to be zero. We now know it has a tiny mass—about one hundred thousand times less than that of the electron.

Finally, in 1956, twenty-six years after Pauli had suggested it, neutrinos were detected in the laboratory. The neutrino has turned out to be essential for understanding the structure of matter on the subatomic level as well as how massive stars end their lives as supernovas.

In the same period that he had to suffer the death of his mother and his own disastrous marriage and divorce, Pauli had managed to come up with a concept of enormous importance in one of those hunches that influence the whole of physics and our perception of the world. Perhaps it was his need to rescue the beauty of quantum mechanics that impelled him to take this imaginative leap. No matter what dramas occurred in his personal life, his mind was always focused on physics.

Pauli in the United States

Nevertheless, Pauli could not suppress his pain forever. He spent the following summer—1931—traveling across the United States, to Pasadena, Chicago, Ann Arbor, and New York, lecturing on his new particle. Oppenheimer and Sommerfeld were among his traveling companions. Prohibition was in force at the time, forbidding the sale of alcohol, which Pauli found exceedingly trying. Ann Arbor, however, was close to the Canadian border and there was plenty of opportunity for smuggling. He wrote to Peierls, "In spite of the opportunity for swimming here I suffer much from the great heat. But under 'dryness' I don't suffer at all."

Indeed he did not. By now he was drinking to excess. At a dinner party in Ann Arbor he fell down an entire flight of stairs. "I broke my shoulder and now must lie in bed until my bones are whole again—very tedious," he wrote. As his right shoulder was broken he could not write on the blackboard. Instead of his usual impenetrable lecture style, he was forced to face his audience, his injured arm supported by a metal rod attached to a ring around his expansive waist, while a colleague wrote up the equations. His audiences were enthralled by the brilliance and clarity of his explanations. He kept the real reason for his handicap a secret. The story that went around was that he had injured his shoulder while swimming. One participant at the physics sessions in Ann Arbor commented that he "now runs around with it stuck up in the air like a traffic cop signalling." Sommerfeld called it an inverse Pauli effect. Later

Pauli commented jokingly that it was the only time in his life he had ever raised his hand in a *"Heil Hitler"* salute.

Instead of returning to Zürich immediately, Pauli spent part of September in a small hotel in Manhattan. Much of the time he kept to himself, except for the occasional meeting with "second-order acquaintances"—friends of friends. (A second-order approximation to solving a problem is less exact than one of first order.) He frequented second-order bars—small speakeasies and out-of-the-way bars rather than the elegant ones that Scherrer had recommended.

Pauli enjoyed America, its people and food—excellent but for the Athenaeum Club at the California Institute of Technology, where he pronounced the food *schlecht* (wretched). He was less enthralled by its puritanical side; at one dinner there "was a prayer instead of coffee and cigars—not to speak at all of alcohol." But on the whole he liked it "better than Europe which seems to me now often tiny and clumsy. . . . It is all very simple and very neat here," he wrote to Wentzel.

That evening, he continued, he was looking forward to going to "a very good bar and drinking many whiskies." For, despite his efforts to conceal it, he was falling deeper and deeper into depression. "With women and me things don't work out at all, and probably never will succeed again," he confessed to Wentzel. "This, I am afraid, I have to live with, but it is not always easy. I am somewhat afraid that in getting older I will feel increasingly lonely. The eternal soliloquy is so tiresome." Alone in his hotel room he signed his letter, "Your old Pauli."

Back in Zürich, he went on a binge of drinking and parties and resumed his life of barroom brawls, smoking, and womanizing. Eventually his bitter quarrels with his colleagues at the ETH came to the attention of the administration, who had to call him in and warn him that his position might be in jeopardy despite his brilliant work.

In front of his colleagues he spoke of his divorce from Käthe in witty, sardonic terms, but his behavior reflected his true desperation. Once again he was living a Jekyll and Hyde existence between two different worlds. To add to all this, his always vivid and dramatic dreams were beginning to seep into his waking life. By the beginning of 1932 he had plummeted to a frightening low point. Despite his hatred of his father, Pauli finally decided to heed his advice: to consult the celebrated psychoanalyst Carl Jung.

The Mephistopheles of Copenhagen

That year, Pauli's neutrino hypothesis was the central topic at the annual Easter conference held at Bohr's institute in Copenhagen. It was also the theme of the spoof that customarily concluded the conference, written that year by Max Delbrück, a twenty-five-year-old physicist who had studied with Born.

He and Pauli were close friends; Pauli addressed him in letters as "Max" and signed them "Wolfgang." Pauli characterized their friendship as a mutual attraction between "two problematic temperaments."

Years later, Pauli reminded Delbrück that "for me personally the history of the neutrino is inseparably connected with your—very *un*successful—flirtation with Eve Curie at the party [at the Nuclear Physics Meeting in Rome, in 1931]. She had a sincere veneration of her old mother, whom she had accompanied to the Rome meeting, but otherwise this icy woman had nothing on her mind other than publicity, newspapers with her name in, etc. Why should she have any interest in you, if there was no chance for her to increase her publicity with your help?"

Delbrück entitled his spoof "Faust in Copenhagen," and modeled it on Goethe's *Faust,* with the luminaries of physics as its characters: God stood for Bohr, and Faust for Ehrenfest (who had said to Pauli, "I like your publications better than I like you"). Mephistopheles stood for the sharp-tongued Pauli, who was also the progenitor of the neutrino (Gretchen). In the spoof, Felix Bloch played God (Bohr) and Léon Rosenfeld, Bohr's assistant at the time, played Mephistopheles (Pauli).

Mephistopheles/Pauli the troublemaker tries to tempt Faust/Ehrenfest by offering him the seductive neutrino—a hypothesis of which the conservative Ehrenfest was famously deeply skeptical. Faust declares that no elementary particle could possibly exist with neither mass nor charge; it is pure madness.

The spoof was wonderfully translated by the physicist, prankster, and frequent visitor to the institute, George Gamow, who had provided delightful cartoons of the characters, including a wickedly accurate depiction of the plump Pauli as Mephistopheles with an infuriating grin and a long tail.

The play opens with Mephistopheles leaping into the midst of a

Pauli as Mephistopheles in George Gamow's caricature.

group of archangels, headed by God, all busy discussing astrophysical matters and how stars shine. "To me the theory's full of sound and fury," he declares. The Lord/Bohr demands,

> But must you interrupt these revels
> Just to complain, you Prince of Devils?
> Does Modern Physics never strike you right?

Mephistopheles replies,

> No, Lord! I pity Physics only for its plight,
> And in my doleful days it pains and sorely grieves me.
> No wonder I complain—but who believes me?

Delbrück perfectly captured the rivalry between Bohr and Pauli, who kept each other at arm's length. Pauli had initially proposed the neutrino specifically to counter Bohr's suggestion that the laws of conservation of energy and momentum held only on average in the case of beta-decay.

The Lord dismisses the neutrino hypothesis in Bohr's much-feared words, saying that it "is very in-ter-est-ing"—incorrect, in other words. Mephistopheles fires back, "What rot you talk today! Be quiet!"

Ehrenfest doubted more than Pauli's neutrino. He also questioned the work on quantum electrodynamics in which Pauli and Heisenberg had been immersed for five years, plagued by the infinite values for the electron's mass and charge, which they could not eliminate no matter how hard they tried. Gamow portrayed the oblong-headed Heisenberg and devilish Pauli as Siamese twins.

Mephistopheles, transformed into a bowler-hatted traveling salesman, tries to sell Faust quantum electrodynamics, the theory formulated "By Heisenberg-Pauli." "No sale!" Faust shouts. Then Mephistopheles offers Faust "something unique"—his neutrino theory. Says Faust,

You'll not seduce me, softly though you speak.
If ever to a theory I should say:
"You are so beautiful!" and "Stay! Oh, stay!"
Then you may chain me up and say goodbye—
Then I'll be glad to crawl away and die.

Ehrenfest famously believed that beauty was for tailors, not scientists.
Finally Gretchen herself, the neutrino, enters singing:

My mass is zero,
My Charge is the same.
You are my hero,
Neutrino's my name. . . .

I am your fate,
And I'm your key.
Closed is the gate
For lack of me.

But Ehrenfest remains unconvinced.

Pauli was not present to see himself lampooned, but he later received a copy of the text complete with Gamow's drawings of him as the mischievous troublemaker of physics. He proudly showed it to visitors. He was obviously delighted to be cast in this role.

In fact he was in Zürich, and his life was about to change.

8

The Dark Hunting Ground
of the Mind

PAULI ALWAYS liked to be well informed and, in preparation for his first meeting with Jung, had no doubt studied several of Jung's books which he kept in his library. Over the course of their work together he read most of the collection. He marked them in pencil: a vertical line for an important passage, two for very important, three for extremely important.

He paid particular attention to *Psychological Types*, the book in which Jung laid out the vocabulary and framework for his analytical psychology. In this he identified two poles of personality—extravert and introvert— and four "functions," thinking versus feeling and intuition versus sensation. *Psychological Types* contains by far the most markings of any of Jung's books in his library. No doubt Pauli was struck by the similarity between Jung's tug-of-war between pairs of complementary functions and Bohr's complementarity principle. Complementarity seemed to be everywhere. Just as it clarified issues in physics for him, perhaps he felt that Jung's familiar-sounding words might provide a key to his inner self.

Pauli marked the following passage with three vertical lines: "Where the persona is intellectual, the soul is quite certainly sentimental. . . . A very feminine woman has a masculine soul, and a very manly man a feminine soul. This opposition is based upon the fact that a man, for instance, is not in all things wholly masculine, but has also certain feminine traits." Pauli was certainly intellectual and equally certainly sentimental, battered as he was by the traumas of his emotional life. He was also a man, and a manly one. But where was his feminine soul? Perhaps Jung would help him discover it.

Jung's description of the introverted-thinking type was an uncannily precise description of Pauli himself:

> His judgment appears cold, obstinate, arbitrary, and inconsiderate; only with difficulty can he persuade himself to admit that what is clear to him may not be equally clear to everyone; [if] he falls among people who cannot understand him, he proceeds to gather further proof of the unfathomable stupidity of man; he may develop into a misanthropic bachelor with a childlike heart; he appears prickly, inaccessible, haughty; [he has] a vague dread of the other sex.

No doubt all of this was on Pauli's mind as he stepped into the entry hall of Jung's house.

Proceeding up a wide turning staircase, he reached the first floor, then, turning right down a hallway, passed the small office Jung sometimes used as a retreat. This was a compact room without any bookcases, immaculately laid out with a desk with three drawers, a small desk lamp, and a rack of pigeon holes for filing papers. Stained-glass windows depicting mythological scenes provided a muted natural light. Seated in his desk chair on a comfortable pillow, Jung would smoke his pipe here while he wrote up reports and articles and replied to correspondence.

In front of Pauli was Jung's spacious library packed with ancient alchemical texts. The floor was covered with oriental rugs. A green-tiled stove kept the room warm in the winter while breezes blowing in from the lake kept it cool in summer. There was a writing desk with a straight-backed chair and a desk lamp opposite the doorway, next to a large window looking out onto Lake Zürich. A couch flanked by two easy chairs occupied the opposite end of the room. Patients could choose either chair, depending on whether they preferred to look at the bookcase or the lake.

Jung in 1930.

Jung used the small room to analyze patients whose problems did not particularly interest him. For those he found emotionally involving, he preferred the library. There he had his alchemical books on hand, added to which, as he put it, the size of the room gave him the mental space for what he termed an out-of-body experience. On these occasions he would "go up and sit on the window, and look down and watch myself, how I am acting, until I see what from the unconscious has caught me and I can deal with it."

It was in the library, sitting on the couch with a table in front of him piled high with books and notes, that Dr. Jung awaited his new patient.

Four years later, Jung described the man who came to see him that day. Pauli was in a shockingly disintegrated state:

He is a highly educated person with an extraordinary development of the intellect, which was, of course, the origin of his trouble; he was just too one-sidedly intellectual and scientific. He has a most remarkable mind and is famous for it. He is no ordinary person. The reason why he consulted me was that he had completely disintegrated on account of this very one-sidedness. It unfortunately happens that such intellectual people pay no attention to their feeling life and so they lose contact with the world that feels, and live in a world that thinks; in a world of thoughts merely. So in all his relations to others and to himself he had lost himself entirely. Finally he took to drink and such nonsense and grew afraid of himself, could not understand how it happened, lost his adaptation, and was always getting into trouble. This is the reason he made up his mind to consult me.

Pauli was a man dominated by intellect, who focused his thinking entirely on the world outside of himself and had almost no awareness of what was going on in his own being—and a famous scientist to boot. For Jung it must have seemed an irresistible opportunity to work with such a person and examine what made him tick, while also trying to help him achieve balance.

Later, in the preface to *Psyche and the Symbol,* Jung phrased it thus:

And what shall we say of a hard-boiled scientific rationalist who produced mandalas in his dreams and in his waking fantasies? He had to consult an alienist, as he was about to lose his reason because he had suddenly become assailed by the most amazing dreams and visions. . . . When the hard-boiled rationalist mentioned above came to consult me for the first time, he was in such a state of panic that not only he but I myself felt the wind blowing over from the lunatic asylum!

"He had completely disintegrated"; "he had lost himself entirely"; "he was about to lose his reason"; Jung "felt the wind blowing over from the lunatic asylum!" Perhaps Jung is exaggerating, but nevertheless his phrases make it clear how desperate Pauli was at this point—with a desperation he could not solve in his working life or communicate to his friends and colleagues in the scientific world. To find a solution he had to step out of that world into Jung's extraordinarily different—and eccentric—universe.

Pauli poured out his troubles—his anger, his loneliness, his drunken brawls, his problems with women, and how he frequently made himself disagreeable to men. His dreams, he said, were full of threes and fours and other matters that seemed to spring out of seventeenth-century science, not modern physics. These dreams and visions were driving him to distraction.

Here was, Jung realized, a young man not only in need of help but also "chock full of archaic material." The problem was how "to get that material absolutely pure, without any influence from" Jung himself. There was only one way. To allow Pauli to speak and dream freely, without any suggestions from Jung, Jung had to keep Pauli at a distance. "Therefore I won't touch it," he wrote.

His solution was rather extraordinary. Instead of treating Pauli himself, initially Jung sent him to Erna Rosenbaum, a young, vivacious Austrian student of his. Rosenbaum had studied medicine in Munich and Berlin and had worked with Jung for a mere nine months before he assigned her Pauli as a patient. Pauli was disappointed at being fobbed off on a student, but Jung gave him no choice.

Pauli wrote to her in his usual laconic manner: "[I contacted] Mr. Jung because of certain neurotic phenomena which are connected with the fact that it is easier for me to achieve academic success than success with women. Since with Mr. Jung rather the contrary is the case, he appeared to me to be quite the appropriate man to treat me medically." But then, Pauli continued, Jung had surprised him by refusing to treat him, sending him instead to her despite the fact that "I am very touchy toward women and slightly distrustful and thus have some hesitations against them. Anyway," he concluded, "I want nothing to be left untried."

Jung later revealed that he had specifically selected a woman as Pauli's analyst because he was convinced that only a woman could draw out a man's thinking from the depths of his unconscious, particularly in the case of a highly creative person such as Pauli. As far as he was concerned, Rosenbaum was the perfect conduit to encourage Pauli to record his dreams. Jung instructed her to play a "passive role," to do no more than provide encouragement and indicate points that Pauli should work out more clearly. "That was enough," wrote Jung. He perceived intuitively that Pauli "had the gift of visualizing things and so he had spontaneous fantasies" as well as dreams. Rosenbaum, in fact, fulfilled her role perfectly.

The day after Pauli's visit, as Jung recalled, she went to see him.

"What sort of man have you sent me?" she demanded. "What's the matter with him? Is he half crazy?"

"What's going on?" Jung asked.

Pauli told her stories with "such emotion that he rolled around on the floor," Rosenbaum replied. "Is he crazy?" she demanded again.

"No, no, he is a German philosopher who is not crazy," Jung replied. Some years later, in a lecture, Jung said that "through this woman [Pauli] had simply realized for the first time that he had a huge amount of emotion about certain things. This he hid from me—I have seen this later again from him. Because in the presence of a man he cannot be inferior. He cannot be inferior!" Jung's intuition was that Pauli could never let down his defenses in front of a man; but with a woman he felt much freer to express himself. His decision to send Pauli to Rosenbaum was the right one.

Pauli saw Rosenbaum regularly for five months, until for reasons which are not clear she left Zürich for Berlin. He continued to record his dreams and tried to analyze them himself. He communicated with her by letter, in which he related his dreams in some detail adding, "I do not envy you for having to read all this."

Meanwhile he was traveling and doing his best to dry out. He stayed at hotels in Portofino and Genoa where no one knew him, and tried to concentrate on a book he was writing on quantum mechanics. But he often became depressed. In September that year, he wrote to Rosenbaum from Zürich complaining about the weather and his intermittent bouts of depression. He mentioned that his sister was in Berlin, and asked for Rosenbaum's phone number, which she did not seem willing to divulge. He added that he hoped to see her again when she returned from Berlin. There is no record of whether he did. In the late 1930s, however, he made several unexplained trips to London. Rosenbaum had moved there to escape the political situation in Germany. There are rumors that their relationship had become more personal. Certainly it sounds as if Pauli had become rather obsessed with her.

In November 1932 Pauli was back in Jung's library. Eight months later they began to meet regularly, on Mondays at noon for an hour or so. Pauli had written up 355 dreams and added another 45 by the time they concluded their sessions a year later. "They contain the most marvelous series of archetypal images," Jung reported ecstatically in a lecture he gave in London some two years later.

Jung often spoke of the dreams "of a great scientist, a very famous young man" in his lectures, but at Pauli's insistence he never revealed his identity. Concerned with his professional reputation, Pauli preferred to keep his sessions with Jung a secret.

Jung first described Pauli's dream sequence at the annual Eranos lectures in 1935 in Ascona, Switzerland. The physicist Markus Fierz, who became Pauli's assistant the following year, claimed that he immediately guessed that the dreamer was Pauli, and others suspected it too. But none of Pauli's colleagues ever revealed anything.

In fact the question of who the "brilliant young scientist" was remained a mystery for fifty years until Charles A. Meier, Jung's successor at the ETH (where Jung had been on the staff since 1933), revealed that Pauli had been in analysis with Rosenbaum. Shortly afterward, Aniela Jaffé, who had been Jung's personal secretary, confirmed that the dreams Jung had often discussed and referred to were indeed Pauli's.

Of Pauli's four hundred dreams, Jung looked at fifty-nine in detail. The ones he chose exemplified the process of what he called individuation. This is a specifically Jungian term referring to the process by which one develops an individual personality. In terms of analysis, individuation is said to have occurred when the patient achieves a balance between the conscious and unconscious. The mark of this is that the patient begins to dream of mandalas—diagrams, usually based on a circle or square with four symbolic objects symmetrically placed.

In the state of individuation the four psychological functions— thinking, feeling, intuition, and sensation—are fully in the conscious and together form an integrated whole. Before he began analysis, the thinking function dominated Pauli's conscious while his feeling function was totally submerged in his unconscious, and his sensation and intuition functions were both partly submerged. He was a totally cerebral personality, out of touch with his feelings. Jung depicted Pauli's psychological state thus:

Jung's theory was that dreams emerge out of the unconscious and therefore offer a means of understanding how it works. Dreams appear when the level of consciousness sinks below the unconscious, a situa-

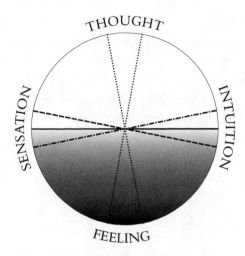

THOUGHT

SENSATION

INTUITION

FEELING

The four consciousness functions in Pauli's case. Thinking, the superior function, occupies the upper half of the circle. Feeling, the inferior function, is in the dark half. The two auxiliary functions, sensation and intuition, are partly in light and partly in dark.

tion most likely to occur during sleep. When we wake up the level of consciousness rises and the world of the unconscious disappears. But dreams can also occur when we are awake. Jung referred to such waking dreams as "visions." Dreams and visions, he wrote, are the two keys to the unconscious.

Jung paid great attention to the imagery in a dream, linking it with images from alchemy, religion, and myth and applying his analytic psychology to seek out archetypes. In this way he hoped to use the opposition between the conscious and the unconscious to enable a patient to meet his "shadow"—his dark side—and to separate it from his anima, the female aspect of the male personality. This would bring about a struggle between opposites—of function types or dream symbols—that would enable the patient to come to terms with the fact that he himself was a combination of light and dark, and good and evil. Thus he could eventually create a balanced personality.

When archetypal symbols, most particularly mandalas, appeared in a dream, it often signaled that a previously disordered conscious was becoming ordered. This, however, did not necessarily mean successful analysis, Jung advised. "There are plenty of lunatics with the most wonderful individuation dreams, and nothing comes of it because there is nobody home," he said.

So what sort of dreams did the tortured scientist have and how did Jung help him work through them? A selection of Pauli's dreams and

Jung's comments on them give a flavor of the process through which Jung marked out a path through his chaotic mental state and helped him work toward individuation.

We cannot know exactly what transpired between Jung and Pauli in the privacy of Jung's study. I have put together the following dialogue on the basis of the descriptions Jung made soon afterward and details from Pauli's biographical materials.

Three women

Pauli dreams he is surrounded by a group of female forms. He hears a voice somewhere inside himself saying: "*First I must get away from Father.*"

Jung's first comment is that the phrase "get away from" needs to be completed by the words "in order to follow the unconscious," as embodied in the seductive female forms. Rising from his chair he walks over to his collection of alchemical texts. He opens a sixteenth-century book. In the image, Pauli's own dream is depicted with amazing precision. It shows the sleeping dreamer, three maidens who, says Jung, signify the unconscious and Hermes—the ancient Greek name for Mercurius, the central figure of alchemy, who moves between the dark and light worlds. In Jung's interpretation of alchemy, Hermes is the intermediary between the conscious and the unconscious.

Jung suggests that the "Father" of the dream is not Pauli's actual father but represents the masculine world of the intellect and of rationality, in opposition to the unconscious. Perhaps the dream indicates that Pauli fears that giving rein to the unconscious will mean sacrificing his intellect, whereas in fact it is a matter of entering an entirely different world with different but equally meaningful experiences. As yet he is not able to attribute to the unconscious its proper reality. Jung adds that Pauli will encounter this problem repeatedly until he can find a way to balance the conscious and the unconscious in his psyche.

In order for the modern scientific world to develop, it has been necessary—in Jungian terms—to relegate the unconscious to a position below the conscious and rationality. In Pauli's case, this marginalizing of the unconscious has been an inevitable consequence of his life as a

The awakening of the sleeping king, shown as a judgment of Paris, with Hermes Trismigestus as psychopomp. (Thomas Aquinas, De alchimia *[MS, 16th century].)*

scientist. It is no accident that the figures in his dream are feminine, for the unconscious is feminine in nature. Like the seductive maidens of mythology and alchemy who appear to lead the unwary traveler astray, Pauli's unconscious is reaching out to him. Pauli can run away if he wants to, but it seems he does not wish to do so. Rather he wants to "get away from Father"—from intellect and rationality which have dominated his life so far.

The serpent Uroboros

A few days later Pauli dreams that he is rooted to the center of a circle formed by a serpent biting its own tail.

Jung reaches down another book, which has a picture of the creature whom alchemists called the Uroboros, a serpent who devours his own

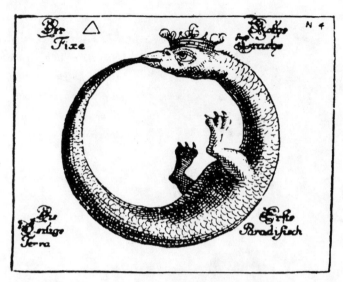

Uroboros symbolizes the process in which the four elements, earth, water, air, and fire, are transformed into each other. (Abraham Eleazar, Uraltes chymisches Werk [18th century].)

tail and gives birth to himself. Uroboros slays and is slain, resurrects and is resurrected, in an eternal and magical transformative process.

The Uroboros symbolizes the eternal circle, the circular process by which the four elements (earth, water, air, and fire) transform into one another. The circular form taken by the Uroboros is also the first hint of the symmetrical form of a mandala, suggesting that change is beginning. The area that the Uroboros encircles is a protected area, the *temenos*, where the dreamer—Pauli—can safely come face to face with his unconscious.

The veiled woman

Then Pauli dreams of a veiled woman.

This is the first time the veiled woman has entered Pauli's dreams. She has done so because the serpent has created a protected area where she can safely appear. Jung tells him she is his feminine side—his anima. The appearance of a person, rather than a symbol, means that the unconscious is stirring. Something has awoken. Pauli's anima will lead him to

*"Atmavictu," totem
carved by Jung in
1920. He claimed that
it reminded him of the
one he had carved as
a boy and that his
unconscious supplied
the name.*

Jung at Lake Zürich, 1920.

Jung in his library in 1946, when he and Pauli resumed their conversations.

Christiani Rosencreutz: Chymische Hochzeit. Anno 1459.

Arcana publicata vilescunt: et gratiam prophanata amittunt.

Strassburg: Lazari Zetzners S. Erben. 1616.

1 Abend vor dem Ostertag, ein grausamer Wind, "das ich nit anders meinte,
dann es würde der Berg, darein mein Häuslein gegraben, vor grosser
Gewalt zerspringen müssen." Jemand berührt ihn am Rücken.
Zupft ihn am Rock — ein "schön herrlich Weibsbild" in blauem Kleid
mit goldenen Stern mit Posaune, Einladung:

3 Heut, heut, heut,
" Ist des Königs Hochzeit,
Bist hierzu geboren,
Von Gott zu Freud erkohren,
Magst auff d. Berge gehen,
Darauff drey Tempel stehen,
Daselbst die Geschicht sehen.

Halt wacht,
Dich selbst betracht,
Wirst du dich nit flissig bad,
Die Hochzeit kan dir schad.
Schad hat, wer hie versaumbt,
Hüt sich, wer ist zu leicht.

Sponsus et Sponsa.

P. 4. Traum vom Thurm: Mit
vielen Menschen gefangen
an Ketten. Plötzlich
Deckel geöffnet. Ein Alter
Eysgrauer Mann spricht
wenn man nicht wollte er-
hielt, so bleibt man gefangen.
Seine Mutter wollte ihnen aber
ein Seil hinunter lassen, wo
sich man dran hängen
könne, um heraus zu —
wird 7 × gezogen. ...
11. ... aus ... göttliche ...
... der ... uralte Johan
der Mutter spricht. (u. A.)

Dem viel vertraut,
Dem gilt's an heut,
Darumb sol et Fewer, ...

13 Erwacht am Trommeln: Rüstet sich zur Reise: Blut rothes Band
kreuzweis über die Achseln, 4 rothe Rosen auf den Hut. ...
14 ... Brot, Salz, u. Wasser. ...
16 Findet eine Inschrift... 4 Wege u. ihre Gefahren...
18 ... Schwarzer Rabe verfolgt Taube. Sie fliegt nach
Süd —, erfolgt ... ist ... auf dem 2. Wege.
19 Ein starker Wind macht die Umkehr unmöglich. ...
tal auf hohem Berge. ... Sonne geht unter. Inschrift Procul hinc,

An excerpt from one of Jung's alchemical
treatises.

1936 Congress at the Niels Bohr Institute, Copenhagen. Front row, left to right, Wolfgang Pauli, Pascual Jordan, Werner Heisenberg, and sixth from the left, Otto Stern; third row, sixth from the left, Paul Dirac; fifth row, second from the left, Victor Weisskopf, and fourth from the left, Hendrik Kramers. Standing at left, Niels Bohr and Léon Rosenfeld.

Max Born tugs Pauli's ear in punishment for sleeping late and missing morning lectures, Hamburg, 1925.

Pauli and Ehrenfest sharing a joke, 1929.

Pauli lecturing on his and Heisenberg's theory of quantum electrodynamics, Copenhagen, 1929.

Pauli on vacation in Pontresna, Switzerland, winter 1931/1932.

Hertha, Pauli's glamorous sister, in 1933.

Sommerfeld (on left) and Pauli, in Geneva, October 1934.

Pauli's father with Franca, 1936.

*Pauli and Franca shortly
after their marriage.*

*Pauli and Wu in Berkeley between
1941 and 1945.*

Scherrer and Pauli, after World War II.

Heisenberg and Pauli in 1957, discussing their unified field theory.

his unconscious and reveal its contents, but he must beware; there may be unpleasant surprises in store. He may find irrationality lurking there.

Jung shows Pauli a picture of veiled women like the woman in his dream moving up and down a staircase, symbolizing the ascent of the soul through the seven spheres of the planets toward the sun, the divine sphere from which the soul originates. (In ancient pre-Copernican astronomy, there were seven planets. Copernicus realized there were only six; one of the seven was the moon.) Perhaps Pauli's dream relates to an initiation rite and moving up the staircase symbolizes the beginning of his transformation into a new person.

Jung also examines the role of woman and of the eternal female in Pauli's personal life. Jung assumed that Pauli must have originally projected his anima onto his mother, as men usually do. No doubt this made Pauli think of his mother, Bertha, who had died six years earlier. The mother symbolizes the source of life—the unconscious, where Pauli's feeling function is hidden. While a man continues to project his anima wholly onto his mother, his feelings too—his Eros—remain identified with her, pushing all other women into the background. This sort of man

Jacob's dream, as depicted by William Blake (19th century).

takes a passive view of life, for he is still in an infantile state. His relationships are passionless, usually restricted to prostitutes.

As Jung says, Pauli's behavior exactly fit this analysis:

> The dreamer repeatedly found himself in the most amazing situations. For instance, he once found himself in the midst of a great row in a restaurant, and a man threatened to throw him out of the window on the first floor.* Then he grew afraid of himself. He did not understand how he got into such a situation. Anyone outside could see very clearly how he stumbled into it. But to himself, he was a victim of circumstances; he had no control over his outer conditions because he was still an embryo suspended in the amniotic fluid where things simply happen. He was a victim of circumstances in this way because he was not related. This is what happens to such a nice boy continuously. He has one affair after another, and is always the victim.

Recognizing himself, Jung adds, Pauli "says: 'What could I do?' like that, like a so-called innocent girl, 'What could I do?' He held my both hands and kissed me." Through his dream work and his intuition, Jung has quickly accessed the depths of Pauli's being and brought some of his most disquieting behavior into the light of day.

Pauli's mother

Soon afterward Pauli dreams of his mother pouring water from one basin into another.

Later, Pauli has a sudden recollection: the other basin had belonged to his sister, Hertha. Perhaps the dream means that Pauli's mother has transferred his anima—his feminine side—to his sister. His mother is superior to him, but his sister is his equal. Thus in this dream he is freeing himself from his mother's domination and also from his infantile attitude toward life.

---

*That is, second floor in America.

Pauli had always kept in touch with Hertha. At the age of seventeen she had left the gymnasium to study acting at the Academy of Dramatic Arts in Vienna. Two years later she made her first stage appearance in Breslau, now the Polish city of Wroclav. She was such a dazzling success that the theater impresario, Max Reinhardt, swept her off to Berlin to join his famous German Theater. There she widened her scope to perform on radio and in film. Pauli often boasted about his glamorous sister and enjoyed visiting her backstage after performances. It was also a way to meet other actresses.

Pauli confesses to Jung that all the women he has ever fallen in love with either looked like Hertha or, like Käthe Deppner, were her friends. But he has never felt close to her. He tells Jung that Hertha was born in his seventh year. Seven is a mystical number: the number of planets on their spheres, the number of days in the week, the number of orifices in the head, the number of voices heard by Moses on Mount Sinai. And seven marks the moment when his anima—his feminine aspect—was born, when a female other than his mother entered his life and he was no longer the center of attention.

Jung predicts that his anima will soon pass from Hertha to an unknown woman mired in his unconscious whom he still confuses with his dark side or shadow. This process had already begun, when Hertha married a fellow actor named Carl Behr in 1929, a union Pauli disapproved of. Jung interprets this in psychological terms. Pauli had been critical of her marriage because it meant she could not carry his anima any more, and he now had to share her with another man. The loss of his mother substitute—Hertha—further added to his troubled state of mind.

To free himself from his mother and Hertha will be a gradual process, says Jung. He will need Jung's help to work out his relationship to the as-yet unknown new woman.

The sun worshipper

A short time later Pauli dreams that an unknown woman is standing on a globe, worshipping the sun.

The unknown woman has appeared at last, says Jung. She is Pauli's

The coniunctio of the sun and the moon. (Salomon Trismosin, Splendor solis *[MS, 1582].)*

anima and he sees her as a sun worshipper because she belongs to the esoteric beliefs of the ancient world. By separating his intellect from his anima, Pauli has buried the anima in this ancient world. In the same way in the modern world, the dominance of rationality, essential for the development of science, has relegated the anima to a backwater in the human mind.

Now that Pauli's anima has appeared, his consciousness is flooded with energy surging up from his unconscious.

The ape-man

Then Pauli dreams that a monstrous ape-man is threatening him with a club. A figure appears and drives the monster away.

Jung shows Pauli an alchemical text written four hundred years earlier, in which there is an image that exactly mirrors the monster in Pauli's dream. "You see, your dream is no secret," Jung tells him. "You are not the victim of a pathological insult and not separated from mankind by an inexplicable psychosis. You are merely ignorant of certain experiences well within the bounds of human knowledge and understanding." Far from being the unique fantasies of a madman, Pauli's dreams are phrased in precisely the same imagery in which humankind has delineated the inner quest—the quest for oneself—over hundreds of years. For Pauli the picture of the ape-man enables him to see "with his own eyes the documentary evidence of his sanity."

There are creatures in the psyche about which we know nothing at all, says Jung. He interprets the figure in Pauli's dream who scares the monster away as Mephistopheles—Pauli's intellect, his rational side.

Pauli has now reached a turning point in his therapy. He has used "active imagination" to reach down into the contents of the unconscious which lie just below the level of consciousness—a method Jung developed from studying the trance states of shamans and medicine men. To do this

A fifteenth-century version of the "wild man." (Codex Urbanus Latinus *[15th century].)*

Pauli has to suspend his critical faculties, to permit emotions, feelings, fantasies, obsessive thoughts, and even waking dream-images to bubble up from the unconscious—a particularly difficult process for a rationalist like him. The danger, warns Jung, is that the patient can become trapped in a world of phantasmagoria.

The perpetual motion machine

A few weeks later Pauli dreams of a pendulum clock ticking on forever without any friction, a perpetual motion machine.

Jung is pleased that Pauli's rational brain has not stepped in and rejected this machine as an impossibility. He interprets it as the second appearance of the eternal circle. Pauli's dream of the serpent Uroboros encircling the dreamer was the first appearance of a circle—a mandala—and the first evidence of a change in Pauli and was quickly followed by the first appearance of the unknown woman, his anima. Similarly this second circle means a step forward in the process.

Three becomes four

Then Pauli dreams that he is with three other people, one of whom is the unknown woman.

Jung interprets the four people as the four functions of the fully rounded personality—thinking, feeling, intuition, and sensation. In his dream, Pauli does not converse with the unknown woman, his anima. She remains in the darkness of his unconscious, for she is his feeling function, which is still submerged. Jung's analysis is that "the feminine nature of the inferior function derives from its contamination with the unconscious as personified by the anima." In other words, the inferior function—feeling—is contaminated by being submerged and therefore close to the unconscious, and this is why it is feminine.

As Pauli opens himself up to allow these different parts of his being to appear, he is also exposing himself to danger. Emerging out of the unconscious, the anima is imbued with tendencies which, when brought to conscious life, may manifest themselves as antisocial behavior. Men normally resist the urgings of their animas, which are often the cause of trouble. But to repress such tendencies could result in the development of a neurosis.

Jung insists that mythology—and its descendent, alchemy—requires the female element to emerge from darkness to become the fourth entity. This will set the stage for the union of irreconcilable opposites—man and woman—symbolizing every primordial opposing pair, such as brother and sister.

To illustrate this, he tells a story. The Babylonian creation myth, the "Enuma Elish," describes a matriarchal world ruled by the goddess Tiamat, the salt water, who represents the unfathomably deep ocean—chaos. Tiamat was murdered by her grandson Marduk in an act of unimaginable violence. The result, says Jung, was a shift in the world's consciousness toward the masculine, casting femininity into the darkness of the unconscious. Pagan and Christian myths, alchemy, and Eastern religions denote odd numbers as masculine. Thus in Christianity the masculine Trinity, three, is also the One. Even numbers, conversely, are feminine. The time has now come to release the feminine unconscious, to create balance by turning three into four.

Pauli is now ready to plunge into the sea of the unconscious. But he still feels an unbearable tension between the conscious and the unconscious, rationality and irrationality.

The square

Then one night Pauli has a terrifying nightmare. People circulate around a square formed by four serpents. As they walk they must let themselves be bitten at each of the four corners by foxes and dogs. Pauli is also bitten. In the center of the square, a ceremony is going on to transform animals into men. Two priests touch a shapeless animal lump with a serpent, transforming it into a human head.

Jung is elated that Pauli has dreamed of a square for the first time. He presumes that it arose from the circle as a result of the four people who appeared in Pauli's earlier dreams. Like the space enclosed by the

Uroboros serpent, Jung says, the square is a temenos, a stage, a protected area where the drama can be played out.

In the dream of the ape-man, Pauli was threatened by a monstrous ape and saved by his intellect in the form of a Mephistopheles figure. Jung interprets this new dream to mean that man, in the prehuman state of his animal ancestors—the shapeless animal lump—is about to be created anew. Pauli is about to be reborn.

The leftward march of the people around the square signifies that Pauli is now focused on the center. He is moving toward centering the psyche, toward individuation, reaching toward his unconscious. Alchemical parallels enter Pauli's dreams, permitting a creative play of images as a way of fusing the apparent irreconcilables of the conscious and the unconscious. This fusion is represented by the alchemical marriage, the coniunctio, the union of opposites: fire and water, man and woman, yang and yin.

Archetypal images depict the alchemical marriage as sexual union. Pauli can now sense the alchemical opposition between three and four,

A temenos—a symbolic city representing the center of the earth, with four protecting walls laid out in a square. (Michael Maier, Viatorium, hoc est, De montibus planetarum septem *[17th century].)*

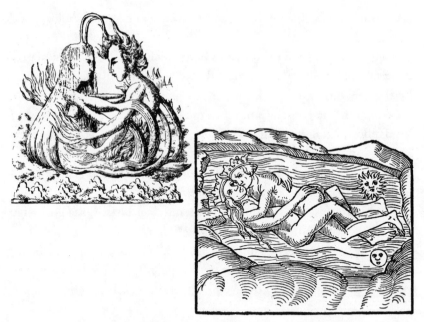

LEFT. *Marriage of water and fire, the union of irreconcilables. Each figure has four hands to symbolize the many different combinations of fours. (Ancient Indian painting;* Nikolaus Müller, Glauben, Wissen und Kunst der alten Hindus *[1882].)* RIGHT. *Psychic union of opposites. (*Rosarium philosophorum *[1550].)*

the trinity and the quaternity. Jung tells Pauli the axiom of Maria Prophetissa, spoken seventeen centuries ago: "One becomes two, two becomes three, and out of the third comes One as the fourth." Pauli's psychological journey—a story of threes and fours—is coming together with his scientific journey.

Coniunctio—the alchemical wedding—is a symbol for the alchemists' eternal quest to create the philosopher's stone or *lapis* through the fusion of the four opposing elements. Pauli's nightmare is an attempt to achieve individuation, like the alchemists' striving for the lapis.

The primal hermaphroditic nature of man and woman—essential to Jung's psyschology in which the female anima exists in man—relates to the four, the quaternary. The *Rosarium philosophorum*, a thirteenth-century alchemical treatise, provides a graphic account of this process. "Make a round circle of man and woman, extract therefrom a quadrangle and from it a triangle. Make the circle round, and you will have the Philosopher's Stone." Thus Jung explains it.

The conscious and unconscious have touched as the alchemical marriage takes place. Now they try to fly apart. But the magic circle traced by the walking figures in Pauli's dream prevents the unconscious from breaking out; the conscious mind takes a stand.

Reflecting on the image of two priests creating a head out of a shapeless mass in Pauli's dream, Jung shows Pauli a fifteenth-century image of God creating Adam from a lump of clay. Perhaps this is the deeper meaning of the image. Pauli is re-creating himself.

In his dream Pauli is bitten by foxes and dogs but this is a good sign, for transformations in the psyche require suffering.

As for the serpents in Pauli's dream, rites of transformation involving serpents are standard archetypes. The serpent appears in Gnostic ceremonies of healing and in their representations of Christ, sometimes on a cross.

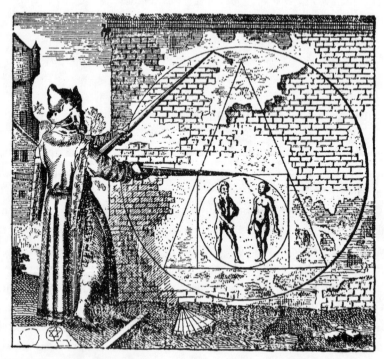

Squaring the circle to make the two sexes one whole. (Michael Maier, Scrutinium chymicum *[1616].)*

God creates Adam from the clay of materia prima. *(Hartmann Schedel,* Das Buch der Chroniken *[1493].)*

Jung sees the square formed by the four serpents as an archetypal ground plan revealing an ordering of the unconscious. But why four?

The basis of alchemy is the reconciliation of opposites. In Jung's psychological theory of types, the least differentiated of the four functions remains in the collective unconscious; in Pauli's case, it is the feeling function. The problem is how to fuse this fourth function with the other three.

When the feeling function emerges into consciousness it releases the Self. Since the inferior function signifies the feminine, the result is

Pagan rites of transformation in the Middle Ages. (Joseph Hammer-Purgstall, Mémoire sur deux coffrets gnostiques du moyen âge *[1835].)*

a wavering between masculine and feminine. The development of the symbols in Pauli's dreams is a sign of the healing process.

Left and right

Some time later Pauli dreams that a group of soldiers armed with anti-quated rifles tries to prevent a revolution in Switzerland by "completely throttling" the left. When he is angry, Pauli tells Jung, he often threatens to "throttle" someone. Jung suggests that the soldiers represent the anti-quated view that left is evil.

To find perfect balance, says Jung, left and right—the unconscious and the conscious—have to be like mirror images. Pauli needs to achieve this, to accept the conscious and the unconscious on an equal footing.

The night club

Shortly afterward Pauli dreams he is in a squalid night club. It is a place where feelings do not count, which makes him feel safe. It's like the old days again. There are a few bedraggled prostitutes there. He argues with the unsavory proprietor about the meaning of left and right. He wants to find symmetry, but he is afraid. He is still suspicious of the left, the unconscious. Angrily he walks out and takes a taxi traveling counter-clockwise around the sides of a square. Back in the night club, he tells the proprietor, "*The left is the mirror-image of the right. Whenever I feel like that, as a mirror-image, I am at one with myself.*" The man replies pensively, "Now that's better." Jung adds that the man has left one thing unsaid: "but still not good enough!"

A man of unpleasant aspect

Two days later Pauli dreams he is sitting at a round table with "a certain man of unpleasant aspect." At the center is a glass filled with a gelatinous mass. The round table, Jung says, suggests wholeness. The man sharing it is Pauli's shadow, his dark side, made up of all the qualities that he and others find so repellent. Pauli's anima is absent. He has finally succeeded in separating his anima from his shadow. His recognition that his shadow is separate from his anima is an enormous step forward. His anima is no longer tainted with moral inferiority. She is finally able fully to assume her role as the mediator between the conscious and the unconscious. At the same time the amorphous mass, or prima materia, comes to life.

He dreams again. The vessel on the table is now a uterus, a symbol which stands for the alchemical vessel in which the chaos of the prima materia is transformed by degrees into the lapis, the Self. It is a moment of creation—the beginning of Pauli's rebirth.

For Jung too the work with Pauli was a journey of discovery, of magical transformations. For both of them it was a way to enter "the no-man's-land between Physics and the Psychology of the Unconscious . . . the most fascinating yet the darkest hunting ground of our times."

9

Mandalas

FROM EARLY ON in Pauli's dream journey, circles had begun to appear as a pattern emerged. The first had been the serpent Uroboros biting its tail, which Jung recognized as a primitive form of mandala. The circle appeared again in a more developed form in Pauli's dream of a perpetual motion machine. His nightmare of people circumambulating a square in which a human head is being created out of an animal mass took place not in a circle but in a square, which Jung interpreted as reflecting the four functions of analytical psychology, the four people of Pauli's dreams, and perhaps also the four quantum numbers that Pauli had discovered.

Jung's description of the philosopher's stone—"Make a round circle of man and woman, extract therefrom a quadrangle and from it a triangle. Make the circle round, and you will have the Philosopher's Stone"—is also a description of the classic Tibetan Buddhist mandala. At the most basic level this is the ground plan of a stupa, a Buddhist temple constructed of a hemispherical mound within a square. Worshippers always circumambulate stupas in a clockwise direction, to the right; leftward motion is believed to be evil (hence the word "sinister"). Jung's interpretation was that the right led toward the conscious and the left toward the

Lamaistic Vajra mandala.
(Commentary by Jung in R.
Wilhelm, The Secret of the
Golden Flower *[1949].)*

unconscious. The picture makes it clear that the circle contains the square, the four points of the compass, and thus all of human life.

As the mandalas in Pauli's dreams became more perfect, Jung took this as an indication that he was moving closer in his psychic journey toward individuation—creating a healthy persona. Pauli's dreaming and drawing of mandalas was a clear sign of a growing balance in the mind between the conscious and the unconscious.

Mandalas appear in cultures across the globe and deep into history—from Palaeolithic rock paintings, the mandalas of ancient Egypt, and medieval mandalas with Christ at the center and the four evangelists in the corners, to the sand paintings of the Navaho and the mandalas that play a key part in the religions of India, Tibet, and the Far East. The mandala can be a circle or a square, but it always consists of four objects symmetrically placed around a center that is the seat and birthplace of the gods. As we have seen, four is an age-old symbol representing, depending on the culture, the four rivers of Paradise, the four directions, the four seasons, and the four elements.

Pythagoras saw the square as the symbol of the soul and in Gnosticism and Hebraism it shared the holiness of the four numbers of the Pythagorean tetraktys. In many religions the square has magical, protective qualities. Thus, for Jung four—not three—was the archetypal foundation of the human psyche. He wrote of how amazing it was that the unconscious should present the same images to people across the globe and across time.

Pauli's first mandalas

Pauli first dreams of a distorted mandala. He tries to make it symmetrical, but fails. The horizontal arm is longer than the vertical one, which Jung interprets as a lack of depth—that the ego still dominates in Pauli's psyche. Each arm of the mandala holds a bowl, which Pauli draws as circles. Each is filled with liquid. One is red, one yellow, and one green, but the fourth is colorless, for the fourth basic color, blue, is missing. (To alchemists the rainbow was made up of four colors corresponding to the four Aristotelian elements—red [fire], yellow [air], green [water] and blue [earth].) The mandala in Pauli's dream is not only distorted but incomplete.

Nevertheless, the fact that Pauli is drawing mandalas is a good sign. In alchemy one of the ways to produce the fusion of opposites in the philosopher's stone and the alchemical wedding is to create images. Pauli

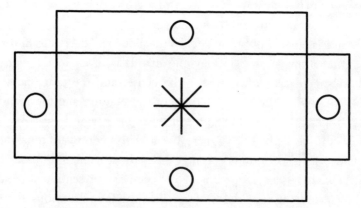

Pauli's distorted mandala.

does precisely this in his fastidious recording of his dreams accompanied by his drawings. In Jung's analytical psychology this shows he is grasping the inner essence of things and depicting their nature as accurately as possible.

In his scientific life Pauli was well aware of the importance of the visual imagination. In the 1930s physics was still in a "period of spiritual and human confusion," as it had been in the 1920s when Pauli had begun to search for a new way to create the true image of the counterintuitive world of the atom. There was no visual image for Pauli's fourth quantum number and this had helped to destroy the visual imagery of Bohr's theory of the atom as a miniature solar system.

The world clock

Then Pauli has a dream that he calls "the great vision—the vision of the world clock." It is an impression of "the most sublime harmony," he tells Jung, and fills him with happiness and peace:

> There is a vertical and a horizontal circle, having a common center. This is a world clock. It is supported by a black bird. The vertical circle is a blue disk with a white border divided into $4 \times 8 = 32$ partitions. A pointer rotates upon it. The horizontal circle consists of four colors. On it stand four little men with pendulums and around it is laid the ring that was once dark and is now golden. . . . The "clock" has three rhythms or pulses:
>
> 1. The small pulse: the pointer on the blue vertical disk advances by 1/32.
> 2. The middle pulse: one complete revolution of the pointer. At the same time the horizontal circle advances by 1/32.
> 3. The great pulse: 32 middle pulses are equal to one revolution of the golden ring.

But what does it all mean? Jung sees in Pauli's world clock a bringing together of all the allusions in his earlier dreams in which its symbols

Pauli's "world clock."

appeared in fragments—the circle, the globe, the square, rotation, the cross, quaternity, and time. In this new dream there is perfect symmetry.

Pauli's world clock has three rhythms or pulses. First there is the small pulse of the vertical ring, which is divided into thirty-two segments. A pointer ticks around them one at a time. When the pointer has passed through all the segments, the middle circle advances by 1/32—the middle pulse. When the middle circle has undergone a complete revolution, "the great pulse occurs"—a single revolution of the golden ring.

The vertical blue circle intersects a circle divided into four parts colored red, green, orange, and blue. On each quarter stands a grotesque dwarf which Jung interprets as a Cabiri, dark gods dating back to ancient Greece who protect navigation. In Jung's analysis they are there to guide Pauli in his journey into the unconscious. Each holds a pendulum. The pulse of the blue circle starts the entire process. Pauli notes the key role of the number thirty-two. Thirty-two is the product of four times eight—4 × 8.

This seems strange. But then Jung points out that thirty-two is a key number in the Kabbalah, signifying wisdom. According to the Kabbalah, thirty-two can be written as the sum of twenty-two (the number of letters in the Hebrew alphabet) and ten (the branches of the Sephirot tree). There are also thirty-two mysterious paths of wisdom.

Pauli has finally dreamed a fully fledged three-dimensional mandala. But Jung is, at first, at a loss as to what it means. What does Pauli mean when he describes his vision as "the most sublime harmony"? It seems to signify wholeness in Pauli's psyche, but why does he state so firmly that he is now at peace with himself? Jung wonders if he is missing some essential clue. Perhaps by "harmony," Pauli means musical harmony, the harmony of the spheres in the sense in which Kepler used the term. Yet the circles are not particularly harmonic; they differ in character and movement.

Another problem is that a mandala always has a sacred object or image at its center. Pauli's has none. The center is nothing but a mathematical point formed by the intersection of the diameters of two circles. It is effectively empty.

Yet Pauli's mandala contains both the masculine Trinity (three pulses) and the feminine quaternity (four colors, four Cabiri), combined to create an alchemical hermaphrodite. Bearing in mind that Pauli is a physicist, Jung speculates on the cosmic significance of this image. Could it be that the mandala symbolizes the four-dimensional source of space-time? But this seems overly scientific. Jung does not have the knowledge to pursue this line of speculation and turns instead to medieval symbolism.

Guillaume's vision

In the last canto of *Les Pélerinages de l'âme* (Pilgrimages of the Soul), the fourteenth-century Norman poet Guillaume de Digulleville describes a vision of paradise made up of forty-nine rotating spheres. (Guillaume's three exquisitely illustrated allegorical poems *Pilgrimage of Human Life*, *Pilgrimage of the Soul*, and *Pilgrimage of Jesus Christ* were to inspire John Bunyan's *Pilgrim's Progress*.) An angel informs Guillaume that these forty-nine spheres represent earthly centuries, not in ordinary time but in eternities. A vast golden heaven surrounds all the spheres. A blue ring, a mere three feet across and half-submerged in the vast golden heaven, glides by. So there are two intersecting systems, one vast and golden, the other small and blue. Guillaume asks the angel why the blue circle is so much smaller than the golden circle of heaven. The angel tells him to look up and he sees the King and Queen of heaven on their thrones.

The angel then explains to Guillaume that the small blue circle is the ecclesiastical calendar and carries the element of time. This very day is the feast day of three saints, the angel says, and begins a rapid discourse on the zodiac. As he tells Guillaume about Pisces, sign of the fishes, he adds that the feast of the twelve fishermen will be celebrated during the sign of Pisces and that all twelve will appear in the Trinity. Guillaume is totally bewildered. What most irks him is that he has never really understood the mystery of the Trinity. The angel launches into a discourse about the three principal colors, green, red, and gold, then stops abruptly and orders Guillaume not to ask any further questions. That is the end of both the canto and the poem.

Guillaume's vision and Pauli's mandala—the quest for the fourth

Guillaume's vision of heaven provides Jung with vital clues both to Pauli's mandala and to his feeling of sublime happiness. In Guillaume's and Pauli's visions the blue circle represents time. In Pauli's mandala it intersects with another of equal diameter, giving a more harmonious fit. The blue circle with its equally spaced segments and ticking hand represents rationalism and thus the masculine Trinity. It drives the circle it intersects, which is segmented into four colors—red, green, gold, and blue, on which stand the four Cabiri. This circle, Jung decides, represents the fourness, the quaternity. The pendulums the Cabiri carry denote the eternal nature of the world clock. The whole mechanism causes the golden circle to rotate. This great circle is no longer dark. In Pauli's psyche, the shadow or dark side has been separated from the anima—his female aspect—which now shines like the sun. No longer buried in the unconscious, it has become enlightened.

The clock on the blue circle sets the entire process in motion. This, says Jung, is because the Trinity is the pulse of the threefold rhythm of the system, which in turn is based on thirty-two, a multiple of four. The circle and the quaternity interpenetrate so that each is contained in the other: three is contained in four.

In a way it was not surprising that Pauli and Guillaume should have had similar visions, in that Pauli too was brought up as a Catholic. Throughout the Middle Ages the problem that hung over the Trinity was that it excluded the feminine. It makes sense that the missing color in Guillaume's dream is blue, the color of the small undeveloped circle. Blue, of course, is the color of Mary's cloak. It is Mary who is missing.

Guillaume was given this clue but he missed it. Instead he saw the King and Queen sitting side by side. But is not Christ the King also in himself the Trinity? Guillaume, a man of the Middle Ages, focused so much on the King that he forgot the Queen. Put the two together— King and Queen, Christ and Mary—and the result is four, a quaternity. Perhaps this was why the angel slipped away before Guillaume started asking awkward questions.

The problem for Guillaume and all the philosophers of the Middle Ages was to find the fourth. Perhaps Pauli's vision provided "a symbolic answer to this age-old question. That is probably the deeper reason why the image of the world clock produced the impression of 'most sublime harmony.' " wrote Jung.

As for the absence at the center of Pauli's mandala, Jung concludes that it is "an abstract, almost mathematical representation of some of the main problems discussed in medieval Christian philosophy." It is only through his knowledge of Guillaume's vision that Jung is able to understand the connection of Pauli's dream with preoccupations going far back into history.

But how could the concept of fourness—the quaternity—arise in the unconscious? The conscious mind could not have put it there. Jung concluded that there must be some element in the psyche expressing itself through the concept of fourness, driving toward the completeness of the individual. Jung emphasizes that the concept of fourness is found in prehistoric artifacts all over the world. It is an archetype often associated with the Creator—though, far from being a proof of God, it proves only "the existence of an archetypal God-image" within human consciousness.

A changed man?

Jung claimed that as a result of his analysis Pauli "became a perfectly normal and reasonable man" and even gave up drinking. He often spoke about the case of the intellectual young scientist as a prime example of the way in which his work on alchemical symbols had helped to shed light on the "development of symbols of the self." It cast light on physics, too.

Two years after he first approached Jung, Pauli wrote to him describing how difficult it had been to cope with the very different but equally repellent parts of his personality before he started analysis:

> The specific threat to my life has been the fact that in the first half of life I swing *from one extreme to the other* (enantiodromia). In the first half of my life I was a cold and cynical devil to other people and a fanatical atheist and intellectual "intriguer." The opposition to that was, on the one hand, a tendency toward being a criminal, a thug (which could have degenerated into me becoming a murderer), and, on the other hand being detached from the world—a totally unintellectual hermit with outbursts of ecstasy and visions.

It shows what an extraordinary degree of self-awareness he had achieved through his dreams and Jung's analysis of them.

A few months later Pauli wrote again to Jung:

> With regard to my own personal destiny, it is true that there are still one or two unresolved problems remaining. Nevertheless, I feel a certain need to get away from dream interpretation and dream analysis, and would like to see what life has to bring me from the outside. A development of my feeling function is, of course, very important to me, but it does seem to me that it cannot emerge solely as the outcome of dream analysis. Having given the matter much thought, I have come to the conclusion that I shall not continue my visits with you for the time being, unless something untoward should arise.

That was the end of Pauli's face-to-face sessions with Jung.

Thanks to Jung, in later years Pauli was somewhat calmer, less acerbic, and less hypercritical, although he was still never seen without a glass of wine in his hand or the occasional martini. Friends guessed that alcohol enabled him to cope with his lifelong bouts of depression.

"The naïve certainty of my former Hamburg days, with which I could easily declare, 'That's all nonsense,' is something I have since rather lost," he wrote to Erich Hecke, a former colleague and friend from those same riotous Hamburg days some years after he stopped his sessions with Jung. Later still, in the middle of a string of critical comments on the work of his former mentor Born, Pauli added wryly to Born, "You will certainly remember old times where I did not have the habit to mix my critical remarks with so much sugar."

As Jung put it, "On a conservative estimate, a third of my cases were really cured, a third considerably improved, and a third not essentially influenced." Pauli fell in the middle.

1 0

The Superior Man Sets
His Life in Order

Franca

PAULI WAS still deep in his sessions with Jung when he happened to go to one of Adolf Guggenbühl's parties in 1933. It was at another of Guggenbühl's famous parties, three years earlier, that he had had his second fateful meeting with Käthe Deppner. On this occasion he was introduced to an elegant and striking young woman named Franziska Bertram.

Born in Munich, Franca was thirty-two, a year younger than Pauli. Always fashionably dressed, she was cultured and well traveled, a woman of determination and strong opinions. Her parents had divorced and she had been brought up by her mother, first in Italy, then in Cairo, where she went to high school. When World War I broke out the family returned to Munich. She moved to Zürich in 1922, where she had been the personal assistant of Friedrich Adler, an eminent Communist politician—famous for having shot the prime minister of the Austro-Hungarian Empire—among several high-level secretarial positions.

Franca moved in high cultural circles and had just ended a relationship with the Swiss author and film writer Kurt Guggenheim. She was still in a fragile state, but was intrigued by Pauli's strange personality. When Guggenbühl suggested that Pauli drive her home, as they both lived on Hadlaubstrasse—Franca at number 17, Pauli at number 47—Pauli replied off-handedly, "I suppose I could take you along." Franca was not impressed.

Pauli was certainly lacking in social graces but nevertheless, despite his apparent coolness, he set out to court her. Perhaps his gauche behavior had simply been shyness. After all, Franca must have been rather intimidating. Shortly afterward, Franca moved in with him. A year later, she recalled, Pauli said abruptly, "Now we marry."

As the great day approached, Pauli maintained his usual air of indifference. But his assistant at the time, Victor Weisskopf, tells a different story. Being Pauli's assistant was a full-time job. It involved grading problems for Pauli's courses as well as being available for discussions with him about his work and keeping him updated on developments in physics. It was always a struggle to obtain his permission to leave Zürich. Late in March that year, Weisskopf with great trepidation asked for one week's leave to go to Copenhagen. "Why?" Pauli demanded impatiently. "I intend to marry and come back with my wife," Weisskopf explained. To Weisskopf's amazement, Pauli replied, "I approve of that, I am going to get married also!"

Pauli and Franca married in London on Sunday, April 4, 1934. Most likely Pauli chose London because he had never been there. Franca's hooded eyes and half smile make her look uncannily like a female version of Pauli.

A couple of weeks later Jung sent Pauli his "best congratulations." Jung had predicted that Pauli's marriage "would constellate the 'dark side of the collective'," meaning that it would bring the good side of otherwise potentially dark archetypes into his consciousness. Pauli, elated, declared Jung was "perfectly correct." To Jung, Pauli described Franca as someone who had "a similar problem of opposites, but the reverse of mine. . . . She fell in love with my shadow side because it secretly made a great impression on her."

Shortly afterward the couple found themselves seated across the table from Jung at yet another of Guggenbühl's dinner parties. Strangely, Jung totally ignored Franca. To make matters worse, Pauli had only just told her that he had previously been married, to Käthe. How could Jung not speak to her when he "was aware that the new marriage could lead to a devastating catastrophe," she later demanded. Pauli reassured her that "Jung knew [from Pauli's dreams] that the binding would be good."

Franca's conclusion was that Jung had ignored her because of "Pauli's decision to marry"; in other words, that Jung had lost Pauli to her. "Pauli, the extremely rational thinker, subjected himself to total dependence on Jung's magical personality," she remembered bitterly. Her distrust of Jung was augmented by her anger that he had sent Pauli to be analyzed by a mere student, Erna Rosenbaum. She insisted that Pauli end his sessions with Jung. Perhaps, in fact, it was she who was jealous of Jung.

Nevertheless, Pauli acquiesced. He ceased dream analysis with Jung. Colleagues at the ETH such as Hermann Weyl thought that Franca had done him a favor.

Nevertheless, Pauli remained somewhat disturbed and insecure. On a skiing trip with Franca that December he panicked that the "earth was shaking under his feet" and screamed that he wanted to "thrash someone." Weisskopf and his wife were skiing nearby and dropped in to see them. Pauli was angry with Weisskopf because he had made an error in a physics paper and was not speaking to him. Weisskopf was eager to get back on speaking terms but Pauli refused to see him. Weisskopf asked Franca to intervene but Pauli had stopped speaking to her too because she had dented their car.

Back in Zürich, Pauli tried to make it up with Weisskopf. "Don't take it too seriously," he said grandly. "Many people have published wrong papers." Then he ruined everything by adding, "But I never did!"

The following year, Erich Hecke wrote to Weyl that he was concerned about Pauli's mental health. He seemed too preoccupied with "dreaming and waking fantasies." Hecke felt sympathetic toward Franca and referred to the "huge piece of work" she had to contend with in her marriage.

In fact Franca contended well. She took care of day-to-day tasks, put up with his cynicism and, all in all, provided a secure home for him. She gave Pauli what he sorely needed—an ordered life in which he could get on with his work. Theirs was an affectionate relationship. The two of them always appeared comfortable with one another.

Over breakfast, Pauli regularly told Franca his dreams and then wrote them down. She recalled that this routine became increasingly important to him as he grew older. To her his dreams were useless exaggerations. After his death, she destroyed all the records of them she could find.

Though Pauli had stopped going to Jung for analysis, the two never ceased corresponding. Franca could not stop his dreams and Pauli continued sending Jung dreams that "perhaps [may be] of some interest to the psychologist." Jung was ecstatic and promised to " 'excavate' [the] ancient and medieval lines that have led to our dream psychology,"—to continue unearthing the mythical and alchemical aspects of Pauli's dreams and working out what they revealed about archetypes. Jung referred not to "my" but to "our dream psychology," a phrase he never used to anyone else. His patient had become a co-worker.

In search of a fusion of physics and psychology

In October 1935 Pauli had a dream in which he was at a physics conference. In his dream he was trying to explain his dreams to colleagues using everyday language but they could not understand. He realized that his dream was all about the need to find a common language that could be understood by both physicists and psychologists. Writing to Jung about it he played with the idea. Perhaps the term "radioactive nucleus," for example, could be interpreted in psychological terms as the Self. Jung declared it an "excellent symbol" for a constellated archetype in the collective unconscious which then made an appearance in individual consciousness and thus encompassed both the unconscious and conscious Selves.

Over the next few years Pauli forged ahead in his research. He worked on crucial problems in physics and maintained a huge correspondence. He pursued infinities in quantum electrodynamics, damning certain of his colleagues' results as deplorable; mulled over the myriad end products of cosmic rays smashing through the earth's atmosphere; delved into the exciting new subject of nuclear physics; and sought a deeper understanding of his greatest discovery, the exclusion principle.

But he never revealed to his scientific colleagues another issue that continued to preoccupy him: the need for a fusion of physics with Jung's analytical psychology in order to understand first the unconscious and then the conscious. Weisskopf recalled that in all the years he knew Pauli, Pauli never once mentioned the topic.

Pauli's dreams and Jung's analyses of them had led Pauli to the rather extraordinary conclusion that "even the most modern physics lends itself to the symbolic representation of psychic processes," he wrote to Jung, adding that there are "deeper spiritual layers that cannot be adequately defined by the conventional concept of time."

In January 1938, Pauli recorded the following dream and illustrated it with a drawing:

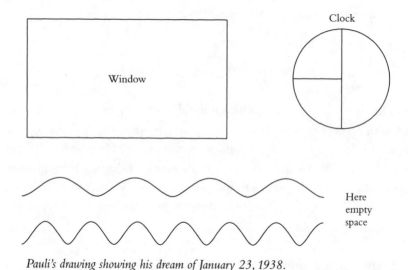

Pauli's drawing showing his dream of January 23, 1938.

In the dream he sees three layers or lines. The top line contains a rectangle, labeled "window," and a circle divided into three sections and labeled "clock." The two other lines are waves with different degrees of oscillation. Pauli and his anima, his female aspect, are both present, but neither can see the time on the clock because it is too far above the lower two levels which he is moving along. So his anima tries to create her own time with what he calls "these odd oscillation symbols," the same as those produced by the dwarves with their pendulum clocks in his "world vision."

Pauli tries to work out the meaning of the dream. He realizes that the rate at which the oscillatory forms vibrate per second must be related to the notion of time. To bring harmony into this system he must find a way to relate all "3 layers to a four-part object (clock)." Once again he is torn between three and four.

Pauli began to notice symbols in his dreams that related to concepts in physics, such as pendulums and time. "In my dreaming and waking fantasies," he informed Jung, "abstract figures are appearing." These included "acoustic rhythms" or "alternating dark and white stripes" like spectral lines or wasps about which Pauli had a severe phobia. "It will become a matter of life and death for me to understand more about the objective (communicable) meaning of these symbols than I do at the moment," he wrote to Jung.

Jung and the rise of Hitler

But no matter how otherworldly they were, in the end neither Jung nor Pauli could ignore the ominous changes in the world around them—to be specific, the rise of Nazism.

Back in 1935 Jung was invited to attend the tercentenary celebrations at Harvard University, scheduled for August 31 to September 18 the following year, as an honored guest.

Behind the scenes, however, there had been a great deal of struggle over whom to invite from Germany—if anyone at all. By now Hitler was in power and professing rabidly racist policies. The situation was particularly complex in the case of Jung who, early on, had become interested in the rise of Nazi Germany from the standpoint of analytical psychology. "Would you have believed that a whole nation of highly intelligent and cultivated people could be seized by the fascinating power of the archetype?" he wrote, adding "the 'blond beast' is stirring in its sleep." He saw this as the archetypal image of Wotan, the mythical warrior king, worshipped before Christianity arrived in central Europe, who had been awakened by the "Hitler movement [which] literally brought the whole of Germany to its feet.... Wotan the wanderer is on the move," he declared, awestruck at witnessing what he saw as his psychology in action.

Jung also expressed ambivalence and downright fear at what he saw happening in Nazi Germany, but nevertheless for him it was an opportunity not to be missed. In 1933 Freud's books had been among those burned in Berlin. Psychoanalysis, so long associated with the Jews, was now banned in Germany. Jung's analytical psychology was the only one allowed in the German cultural scene. Jung wrote several tracts in praise of National Socialism (Nazism) and in condemnation of the Jews, making statements no doubt intended to curry favor with the authorities, such as "The 'Aryan' unconscious has a higher potential than the Jewish." Condemning Freud's psychology, he wrote sternly, "it has been a grave error in medical psychology up till now to apply Jewish categories—which are not even binding on all Jews—indiscriminately to Germanic and Slavic Christendom."

Jung cemented his control over psychoanalysis in Germany by becoming president of the International General Medical Society for Psychotherapy as well as editor of its journal *Zentralblatt für Psyschoanalyse*. But he made sure that the organization's rules were vague enough that they banned Jews only from the German section.

Rumors even reached the Harvard tercentenary selection committee that Jung was "The Mephistopheles in the Nazi drama"—that is, that he was the eminent grise behind the Nazis—and that he was amused at the Nazi's treatment of "Freud's brethren"—the Jews. After many acrimonious sessions the committee decided to invite Jung, placing academic unity over political concerns. It was a decision for which Harvard would later be criticized.

Pierre Janet from France and Jean Piaget from Switzerland were among other psychologists who were invited. Freud, then eighty years old, was not invited. The tercentenary committee decided he was too old to attend and would probably decline.

Einstein declined and Bohr decided not to attend. Heisenberg had also been invited and had accepted but was forced to withdraw. In fact, the German government had granted him permission to travel, but his military obligations and his need to reply to attacks on science in the press required him to stay in Germany. Pauli was not invited. Either the Harvard University scientists did not consider he had the required stature or perhaps they wanted to invite only physicists who were Nobel laureates.

The ceremonial sessions were attended by a distinguished audience of 17,000 who took their seats in Harvard's inner sanctum, Harvard Yard, where the fledgling Continental Army had drilled in 1776 and George Washington had been headquartered. Most of the major figures of European academia were there. Sixty-seven thousand people formally registered to attend the proceedings.

The featured speaker was President Franklin Delano Roosevelt, who was then running for his second term in office. Jung's opinion of him made headlines in *The New York Times*: " 'Roosevelt Great,' Is Jung's Analysis." He thought less about Roosevelt's wife, Eleanor: "the nightmare on the way to being dreamt," he was quoted as saying.

James Conant, Harvard's dynamic young president, read out Jung's official citation for his honorary degree thus: "Doctor of Science. A philosopher who has examined the unconscious mind, a mental physician whose wisdom and understanding have brought relief to many in distress."

Jung's lecture, on the morning of September 7, drew the largest crowd of all the seminars given. He spoke on "Psychological Factors Determining Human Behavior," of how the "human psyche lives in indissoluble union with the body." Afterward there was a spellbinding conversation between Jung and the American modernist poet Charles Olsen on mandala imagery in Herman Melville's novel *Moby-Dick*.

The Jungs stayed in Milton, Massachusetts, at the home of G. Stanley Cobb, an eminent medical researcher. As was European custom, every evening Jung left his shoes outside their bedroom door to be shined, apparently unaware that they did not have live-in help. So as not to embarrass his guest, Cobb shined them himself.

As part of the climax of the splendid celebrations, on September 17 there was a spectacular fireworks display on the Charles River. Half a million spectators lined the banks.

After the celebrations at Harvard Jung gave a series of lectures at Bailey Island (Maine), New York City, and Yale University, where he spoke about how he had treated a brilliant but troubled scientist. Back home he got down to work on his "long overdue" book on alchemy.

According to one (unsubstantiated) story Jung shortly afterward interrupted his work for an undercover assignment. Josef Goebbels, Nazi minister of propaganda, summoned him to Berlin to attend official ceremonies with Hitler, Hermann Göring, commander of the German Air Force, and Heinrich Himmler, head of the feared SS. Jung's task was to assess whether they were crazy. Presumably, if so there would have been a coup. According to the story Jung quickly realized they were all madmen, and, fearing for his life, left immediately.

Germany heads for war

Shortly afterward a very serious problem arose for Pauli—the question of his status in Switzerland. Germany annexed Austria in March 1938 and as a result Pauli's Austrian passport became a German one. He immediately applied for Swiss citizenship but his application was refused. There were problems with residency requirements, added to which Pauli's command of the Swiss-German dialect was not good. The mayor of Zürich informed him that his residency requirement would be fulfilled in spring 1940, after which he should reapply for Swiss citizenship.

So Pauli ended up back at the German consulate in Switzerland. After a cursory examination of his family history, the officials there declared him half Aryan, qualifying him for a straightforward German passport without the "J" stamp (meaning "Jewish"). His German passport was issued in November 1938 and was valid for two years. In Jewish tradition being Jewish is passed through the mother and so Pauli actually was not Jewish. But in German terms he was, because he had Jewish ancestry through his father's family. In fact, under the grotesque arithmetic of Nazi racial theory, Pauli was 75 percent Jewish. As well as his father being Jewish, his mother's father was too. If the Germans were to occupy Switzerland his passport would receive the dreaded "J" stamp, which would mean almost certain death.

As Pauli put it in his inimitable English to Frank Aydelotte, the director of the Institute for Advanced Study in Princeton, New Jersey:

> By the fact that Switzerland didn't make possible my naturalization in
> the moment of the annexation of Austria by Germany I was forced
> to accept a German passport. The German consulate counted me

to the half-Aryans without further examination and so I got a non-Jewish (that means without *J*) passport. Actually I suppose I am after German law 75 percent Jewish. This would mean that in the case of German occupation of Switzerland I would be really menaced and treated as a Jew.

The following year Arthur Rohn, president of the ETH, suggested that Pauli apply again for Swiss citizenship. He did so in December 1939 but heard nothing for six months.

By the end of May 1940 Germany had Switzerland surrounded. Pauli acted quickly. He had been invited to the Institute for Advanced Study as a visiting professor for the winter term 1940–1941. He immediately arranged visas for Franca and himself to travel to the United States. He also pressed Rohn to resolve his citizenship case.

Rohn warned the head of the police division, Dr. Heinrich Rothmund, that the eminent Professor Pauli could be lost to the United States if Swiss citizenship was not granted soon. Pauli also wrote to Rothmund about the delay in processing his application and his attitude toward Germany's annexation of Austria. He received a negative reply.

In a more detailed letter to Rohn, Rothmund declared that Pauli's disapproval of the political situation in Austria and his desire to rid himself of his German citizenship, which he had never wanted in the first place, were not grounds for accepting him as a Swiss citizen. The decisive factor, he added "follows from [Pauli's] characterization, reflected in one of the present police reports from Zürich, given by a closer colleague [regarding] his fitness for naturalization." In other words, as Charles Enz, Pauli's last assistant, wrote, "Pauli's difficulty was due to a colleague!" Pauli was well aware of the animosity of several colleagues to him due to his being a Jew. One, it seemed, had written antagonistic comments about him.

Rohn questioned the police report, even calling for support from Pauli's close friend and colleague Paul Scherrer and the ubiquitous Adolf Guggenbühl, but to no avail.

By now the Swiss authorities were well aware of German expansionist ambitions. Granting a famous Jewish scientist citizenship would not be a politic move. It was only on their second attempt that Pauli and

Franca Pauli wrote of this 1940 passport photograph, "To my opinion, it is the best existing photo of W. Pauli."

Franca managed to make it to the United States after an arduous and sometimes nail-biting journey. They traveled by train through southern France, then across to Barcelona and Lisbon, from where they took a ship to New York, arriving on August 24, 1940. During the journey Pauli lost his nerve several times. Franca had to argue with him fiercely to persuade him to push on through Portugal. Just before he left, Pauli wrote a letter to Jung, concluding in all sincerity, "With my best wishes to you in this difficult time."

Unknown to Pauli, just a few weeks later his sister Hertha took a similar escape route. Her journey was even more harrowing. She had had to leave Berlin in 1933 after the Nazis began to suppress the arts. She went back to Vienna where she founded a literary agency, did some journalism, and began writing novels—all very much in the footsteps of her mother. She arrived with her lover Odön von Horváth, a Hungarian author of political plays that lampooned the Nazis. Hertha had fallen madly in love with him in 1932 and divorced Carl Behr. The two fled Berlin together. In

Vienna they were the toast of the town. His plays were highly acclaimed and along with her beauty and talent as an actress, writer, and sometimes painter they gained easy access to the city's vibrant intellectual world. Her first breakthrough book was the biography of Baroness Bertha von Suttner, the first woman Nobel laureate, awarded the Peace Prize in 1905 for her pacifist activities. Hertha also painted a portrait of her that bears a touching resemblance to her mother, both a pacifist and a journalist. In fact, Hertha's mother had been a close friend of von Suttner's.

When Germany annexed Austria in 1938, the couple fled to Paris where Hertha worked in publishing and continued writing novels. Not long after they arrived, the couple were caught in a violent rainstorm on the Champs Elysées. In a freak accident a branch of a tree fell on von Horváth, killing him instantly.

Two years passed. As the Germans invaded France, Hertha headed south to Marseilles, braving air attacks on refugee columns, German tanks, and the constant threat of arrest by the Vichy police. With the help of the legendary Varian Fry, the American Schindler, she managed to cross the Pyrenees into Spain and from there made her way to Lisbon. After obtaining an "emergency rescue visa" from the International Rescue Committee she arrived in New York City in September 1940. A year later Hertha was in Hollywood writing for Metro-Goldwyn-Mayer studios. In 1942 she returned to New York City where she discovered her talent for writing books for children, usually with Catholic themes. She had always been convinced her escape from France was a miracle, which she attributed to the fact she had passed through the town of Lourdes.

The war years

Arriving in Princeton in 1940, Pauli spent his time working with Einstein on general relativity and continuing his prewar research. He also became close friends with the art historian and Kepler expert Erwin Panofsky, whom he had first met in Hamburg in 1928. Pauli quickly adapted to his new life. He bought a car and drove cross-country to visit colleagues and give lectures on his work. But this was also a trying time. Pauli's German passport meant that he was stranded. At the institute he had difficulties finding funds to extend his stay, which had initially been planned for only one year.

In Zürich, officials and students at the ETH wrote demanding that Pauli return by the end of 1942. Otherwise, they said, his position would be in jeopardy. They regarded his leaving the country as a defection. Colleagues who were jealous of Pauli or had anti-Semitic feelings took the opportunity to vent their anger openly.

In the files of the ETH there is a letter that Paul Scherrer, supposedly Pauli's friend and colleague, wrote to Arthur Rohn, the president of the ETH, in October 1941, saying he opposed allowing Pauli to continue his leave of absence. "Mr. Pauli is naturally having a very good time in the United States; but his productivity has suffered very much—as for all émigré physicists," he wrote. It seems likely that the person who wrote to the police chief Heinrich Rothmund, advising him not to accept Pauli's request for naturalization, was Scherrer and that Scherrer was probably causing Pauli further difficulties in maintaining his professorship at the ETH. He seems to have been more concerned about the future of the physics department in the event of a German invasion than the safety of the man he pretended was his friend.

To the end of his life, Pauli never knew of Scherrer's treachery. What had happened to their previous friendship? Franca always noticed that at ETH functions and in group photographs Scherrer stole the limelight from Pauli. Scherrer's manic need for self-publicity was well known and Pauli used to joke about it, rating people's self-importance in "Scherrer Units." Perhaps he felt he had to support Franca's distrust of the academic hierarchy and drew away from Scherrer. As a result Scherrer began to resent what he saw as a lack of support from the department's most important physicist.

Pauli replied to President Rohn that the ETH officials who ordered him to return were ignoring "the practical impossibility of the journey for me" and threatened to take legal action against the ETH. In response President Rohn hastened to secure Pauli's position until 1948, the end of his second ten-year contract.

Pauli sometimes complained of being lonely at Princeton. "The past years have been rather lonesome, particularly '42 and '43," he wrote to a former postdoctoral student, Hendrik Casimir, in Holland. His one-time colleagues from Europe were now at Los Alamos, working on the Manhattan Project to develop an atomic bomb. Pauli was not, as he later made clear, a great enthusiast of the bomb project. But he was running low on funds and offered his services to the director of the project who happened to be one of his first postdoctoral students, J. Robert Oppenheimer. Oppenheimer rejected his offer. He advised Pauli that he was better off doing pure research, setting an example for physicists who because of "legal complications cannot work on military problems."

Pauli, in any case, was not a specialist in nuclear physics, nor had he ever been a team player. His brilliance was in research at the highest theoretical level, whereas what was required at Los Alamos was very much applied research.

All the same, it is striking that Oppenheimer should have turned down such a distinguished scientist. Perhaps the Pauli effect was on Oppenheimer's mind? After all, there was plenty of delicate machinery, not to mention powerful explosives, at the site.

Pauli inadvertently almost played a part in the war effort. In 1942, his old friend Gregor Wentzel wrote to tell him that Scherrer had invited Heisenberg to give a lecture at the ETH. It was the first time during the war years that Heisenberg had left occupied Europe. Pauli passed the information on to Weisskopf as a bit of friendly gossip. Weisskopf was working on the Manhattan Project and knew that Heisenberg was involved in the German atomic bomb program. He immediately hatched a plan, in which he too would have played a part, to have Heisenberg kidnapped. He passed the plot on to Oppenheimer who passed it to the military. But in the end nothing came of it.

Hertha, meanwhile, who was in New York, had discovered that her older brother was not just an important scientist but a famous one, and that he was near New York. Life was not going well for her and she needed financial and personal help. From time to time, she took to visiting the Paulis. Franca had misgivings about her. Possibly it was connected with Hertha's drive to have a career, while Franca devoted her life to her husband. Or perhaps it was Franca's innate jealousy of any female acquaintance of Wolfgang's, even his sister—perhaps with good reason.

Agent 488

Pauli had the choice to opt out of the war, but Jung could not. As war loomed, Zürich became a nest of espionage and counter-espionage. In neutral Switzerland people moved freely about while keeping constant watch on each other. Parties, social gatherings, and universities were all potential places for exchanging information, finding out who was an agent and for what side, or trying to be a double-agent. Double-crossings were not uncommon, sometimes with dire consequences.

Ordinary citizens had to cope with food and fuel shortages. Like everyone else the Jungs dug up their landscaped lawns to grow vegetables. As a family of means they were able to come up with enough food, tobacco, and wine to maintain at least a vestige of their opulent prewar lifestyle.

Then Allen Dulles arrived, sent by Colonel William J. ("Wild Bill") Donovan, head of the American Office of Strategic Services (OSS), the forerunner of the CIA, to establish a listening post in Switzerland. Dulles's official title was Special Assistant to the American Ambassador in Bern. Before the war Dulles had been a successful Wall Street lawyer. He also had extensive experience in intelligence affairs from World War I. Just hours after he had slipped into Switzerland the Germans invaded Vichy France and closed the French border. Dulles would have to work with whomever he could recruit locally. Meanwhile there was two-way traffic of German spies across the German border with northern Switzerland, aided by a sympathetic population.

One of Dulles's earliest recruits was Mary Bancroft, a thirty-six-year-old American and a well-known socialite. She was famous for her affairs and also for her loose tongue.

Dulles quickly added her to his list of lovers. He impressed upon her the seriousness of her task as a go-between, gathering information from Germans working for the OSS as well as advising him on who was who in Zürich. He cautioned that if she talked too much lives could be lost.

Bancroft knew Jung socially and mentioned him to Dulles in her reports. Aware of Jung's reputation as a Nazi sympathizer, Dulles had him investigated and concluded that the allegations were untrue. The two men met and were impressed with each other. Perhaps Jung was intrigued at the prospect of folding together espionage and psychology.

On Dulles's suggestion, Jung embarked on a series of psychological profiles of Nazi leaders. "It is Jung's belief that Hitler will take recourse to desperate measures up to the end, but he doesn't exclude the possibility of suicide in a desperate moment," Dulles wrote. It turned out to be an accurate prediction.

Dulles considered Jung's profiles dependable and referred to him as Agent 488 in his despatches to the OSS offices in Washington. Jung may also have given him information obtained from patients.

Bancroft had also started analysis with Jung, to bolster her confidence in the spying game. As part of their sessions Jung advised her on how best to question someone based on psychological type, as well as how to apply analytical psychology to the speeches of the top Nazis.

There is a story about Bancroft's unconventional way of communicating with an important German contact. Telephones could be tapped, so this method of communication was used only with the greatest care. Bancroft claimed that when she needed to speak to her contact she used telepathy, willing him to call her. Minutes later he phoned saying, "I just got your message to call."

Dulles was incredulous. "I wish you'd stop this nonsense! I don't want to go down in history as a footnote to a case of Jung's!" he said. But Jung was interested in telepathy and asked her to keep records of how long she spent willing him to call and how long it took him to respond.

Whether true or not, that the story is told at all is evidence of Jung's involvement with intelligence activities in Zürich.

So did Jung have Nazi sympathies or not? The judgment of history is still out. It is difficult to weigh the anti-Semitic opinions he expressed, supporting the Nazi line, against his comments about the dark side of Nazism, though these were never as strongly put during the war. Was his ambivalence an attempt to play it safe? In fact, throughout his life he made anti-Semitic comments. In 1918 he declared that Jews were so overcivilized that they no longer possessed that essential dark Germanic quality—being a pure barbarian brimming with creative potential of the greatest complexity.

He wrote at some length of Freud's psychoanalysis as a Jewish doctrine and described how its reduction of everything mental to material beginnings based on primitive sexual wishes as an oversimplification unsuitable for application to the complex German mentality. He had voiced similar opinions even earlier. In 1897, when he was a medical student at the University of Basel, he spoke to a Swiss student fraternity where he remarked, repeating the then-current prejudice against Jews, that they were materialists who robbed science and culture of their spiritual foundations.

Jung was a man of his times, typical of the Northern Swiss culture, a region that remained neutral yet was sympathetic to the Nazis. But as early as 1934 he realized that he may have overstepped the mark. "I have fallen afoul of contemporary history," he wrote. Yet he persisted.

Many years later, in 1947, Jung invited Gershom Scholem, a well-known Israeli scholar of Jewish mysticism, to lecture at the annual Eranos Conference in Ascona, Switzerland. Aware of the rumors that Jung had sympathized with the Nazis, Scholem asked the highly respected Rabbi Leo Baeck for advice. Baeck had visited Zürich shortly after being released from the concentration camp at Theresienstadt in Czechoslovakia, where he had been one of the camp's spiritual leaders. At that time he had refused Jung's invitation to visit him at home. Jung was insistent and came to Baeck's hotel where they talked for two hours. Defending his stance, Jung spoke of the wartime conditions in which it had not been clear how long the Nazis would be in power, that things might get better, and that to survive it was best to play along with them. Then Jung said, "Well, I slipped up." It was the closest he ever came to an admission of guilt. This satisfied Baeck and they parted as colleagues. Having heard this story, Scholem accepted Jung's invitation and stayed two weeks at his house.

Pauli wins the Nobel Prize

For Pauli 1945 was a momentous year. At the suggestion of Einstein and the mathematician Hermann Weyl, he was offered a permanent position at the Institute for Advanced Study and also at Columbia University. Then came the greatest honor of all: He was awarded the Nobel Prize for his discovery of the exclusion principle.

During a dinner in Princeton in his honor, Einstein gave an impromptu address in which he spoke of Pauli as his successor. Pauli was visibly moved. Panofsky also spoke highly of his friend's knowledge of Kepler and his period.

He recalled their first meeting, in 1928 or 1929, in Hamburg, where they had been introduced by a mutual friend over lunch at an outdoor restaurant. For Panofsky it was an unforgettable occasion on many levels, one being that it provided him with a personal experience of the famous Pauli effect. After the meal, when the three stood up, Panofsky and the friend discovered that the two of them—but not Pauli—had been sitting in whipped cream for the whole three hours. He added two more stories of the Pauli effect. On one occasion "two dignified-looking ladies simultaneously and symmetrically collapsed with their chairs on either side of Pauli" as he took his seat in a lecture hall. On another, Pauli was on a train when, unknown to him, the rear cars decoupled and were left behind while he proceeded to his destination in one of the front cars.

The Pauli effect was surely, Panofsky concluded, based on the Pauli exclusion principle in that whenever Pauli appeared, catastrophes occurred to animate and inanimate objects in his vicinity—but always "*excluding* Pauli himself."

In photographs Pauli is smiling and relaxed. His great discovery had finally been recognized.

In January 1946 Pauli was granted U.S. citizenship. With job offers at Columbia and the Institute for Advanced Study he could easily have stayed in the United States forever—as many scientists, such as Einstein, chose to do. But in fact he decided to return to Zürich and the ETH. It was not so much that he pined for Switzerland: "For me, of course, it is not possible to consider myself as belonging to a single country (that would contradict the whole course of my life). I feel, however, that I am European," he wrote to Casimir. He went on, "I know how bad the material situation in Europe is, and it is true that the material side of life is very well and undisturbed here. I cannot say the same about the spiritual situation."

He was more explicit about what he meant by the "spiritual situation" in a letter to his old friend from his earliest visits at Bohr's Institute in Copenhagen, Oskar Klein, "I am a bit concerned (though not surprised) on this new instrument of murder, the 'atomic bomb'. Although your first hope, that it will shorten the Japanese war, has been fulfilled, I am very skeptical about your other hope, that it will never more be used in any war! I feel that our profession will be discredited among decent feeling persons if the production of this new instrument of murder will not soon be brought under international control."

Pauli never regretted not having taken part in the Manhattan Project to develop the atomic bomb. As he saw it, science in the United States was becoming nothing more than an arm of the military: "As in Austria during the First World War, in this year in the U.S.A. I suddenly had the feeling that I was placed in a 'criminal' atmosphere—and this at the time when those 'A-bombs' were dropped," he wrote scathingly. It has even been said that he once referred to the American scientists who worked on the bomb as "gangsters." So clearly he didn't feel at home in the United States.

Friends say, however, that he simply missed his home in Zollikon, outside Zürich.

Thus it was that Pauli returned to Zürich and the ETH in July 1946. Meeting him again after six years, President Rohn found him totally different from the arrogant character he had been when he left. Pauli declared he wanted to put all the difficulties he had had at the ETH behind him. What had hurt him most, he said, was being judged unworthy of being a Swiss citizen and a professor at the ETH. Nevertheless it took another three years before Pauli was finally naturalized.

He had also missed seeing Carl Jung.

Dreams of Kepler

Once he was settled in Zürich, Pauli quickly got back in touch with Jung and sent him some dreams.

One of his first dreams, which he sent to Jung in October 1946, was about a "blond" man. In the dream, Pauli is reading an ancient book about the Inquisition and how it persecuted disciples of Copernicus, Galileo, and Giordano Bruno, and also about Kepler's image of the sun as a concrete symbol of the unvisualizable Trinity. The blond man tells Pauli that "*The men whose wives have objectified rotation are being tried.*" Then Pauli is in the courtroom with them. His wife is not among them and he wants to send a note to her. The blond man tells him that not even the judges understand what rotation means.

The blond man then says he is seeking a "neutral language" that transcends terms such as "physical" or "psychic."

Pauli kisses his wife goodnight and tells her how sorry he feels for the accused. He weeps. The blond man says to him with a smile, "Now you've got the first key in your hand."

Shaken, Pauli awakens. The essence of the dream, he thinks, is that men have lost touch with their animas—their female aspect, that is, their wives, for their wives, being cut off from the world of science, cannot understand the scientific term "rotation." But what does this have to do with ancient science and with Kepler? Thinking through the problem, Pauli realizes that Kepler too did not fully understand "rotation." Kepler's image of the Trinity as a sphere is also a mandala. But, in the Jungian sense of the term "mandala," it is incomplete in that it is made up of three, not four, elements.

Kepler's image of the creation of the universe is a straight line emanating from the center, from God, like a ray of light emanating from the sun. Pauli's analysis is that this line snags the surface of the sphere and as a result Kepler's mandala is static and cannot rotate. It cannot be a true mandala until it is completed by the fourth element, the anima. This is why in Pauli's dream his wife is absent in the court scene.

Beginning with Kepler, Pauli realizes, modern scientists deliberately excluded the anima (in the Jungian sense of the female aspect of their psyche) as they tried to mechanize the world, partially guided, perhaps, by the image of the Trinity, which they saw in the three dimensions of space. Fludd recognized that modern science's emphasis on inert matter relegated human feeling to the depths of the unconscious. It is when Pauli weeps in his dream, expressing feeling, that the blond man tells him he has found the "first key." Pauli recognizes Kepler and Fludd as opposing psychological types—Kepler the thinking type and Fludd the feeling type. Thus his knowledge of Jungian psychology has revealed to him the limitations of modern science.

Kepler, he thinks, saw the soul "almost as a mathematically describable system of resonators"—like Bohr's virtual oscillators—rather than an entity that could be visualized. Fludd, conversely, focused on four, not three, and used drawings to communicate his beliefs.

It was as he was thinking through this dream that Pauli decided to look more deeply into Kepler and his work. Delighted with Pauli's plan, Jung gave him alchemical literature, as did Panofsky. Pauli also corresponded with his one-time assistant Markus Fierz. Fierz had studied Newton, who was born twelve years after the death of Kepler, and pointed out that his concepts of space and time were saturated with religion; to Newton both space and time were relative to God.

What of Kepler's era, Pauli wondered, when space and time had not yet been elevated to such heavenly heights? He was eager to go back to the moment when mysticism and alchemy clashed with the new rational scientific thinking. He suspected that this collision still went on in "a higher level in the unconscious of modern man."

Early in 1948 Pauli gave two lectures on Kepler and Fludd at the Psychological Club in Zürich. Jung was in the audience. In his lectures Pauli queried the relationship between sense perceptions and the abstract thinking necessary to understand the world around us. How do we generate knowledge from the sense impressions that bombard us? Sensations enter our minds and knowledge emerges. But what happens in between?

We could argue that we have nothing in our minds with which to organize incoming sense perceptions and stumble about learning from experience. But in that case how do we arrive at an exact science such as mathematics from the results of inexact measurements? The alternative is to assume that we are born with certain organizing principles already existing in our minds. Pauli argued that it is archetypes that function "as the long sought-for bridge between the sense perceptions and the ideas and are, accordingly, a necessary presupposition even for evolving a scientific theory of nature." They are, in other words, catalysts for creativity.

A month after Pauli's second lecture, the C. J. Jung Institute opened in Zürich. It was to be the base for a multidisciplinary approach toward understanding the unconscious, which would require forging a link between psychology and physics.

In his speech at the opening ceremony, Jung took particular pleasure in drawing attention to Pauli's work in examining this problem "from the standpoint of the formation of scientific theories and their archetypal foundations."

Pauli, of course, attended and once again his presence had a devastating effect on a material object. In this case it was not a piece of scientific equipment that broke down but a vase that overturned, spilling water all over the ground. Pauli wrote gleefully to Jung about "that amusing 'Pauli effect'." Inspired by Jung's lecture on the importance of linking psychology and physics, he wrote up his own thoughts on the subject in an essay entitled, "Modern Examples of 'Background Physics'."

Dreams of physics

Starting from around 1935 Pauli had occasionally had dreams and fantasies in which "terms and concepts from physics appeared in a quantitative and figurative—i.e., symbolic sense." He called this "background physics." At first he dismissed it as personal idiosyncrasy and was reluctant to discuss it with psychologists because of the physics terminology involved. But then he was struck by the similarity of the symbols in these dreams with the images he came across in seventeenth-century treatises like Kepler's, written at a time when "scientific terms and concepts were still relatively undeveloped."

When he looked into it, he discovered that people who knew nothing of science often created similar images. From this he concluded that his dreams were not, after all, meaningless or arbitrary. It seemed to be proof that " 'background physics' is of an archetypal nature." Because physics and psychology are complementary, he was certain that there is "an equally valid way that must lead the psychologist 'from behind' (namely, through investigating archetypes) into the world of physics." In other words, the prevalence of these symbols seemed to provide firm evidence that the symbols of atomic physics derived from archetypes.

Pauli gave as an example of background physics "a motif that occurs regularly in my dreams": the fine structure of spectral lines. What he was looking for was the underlying meaning of these dreams, their "second meaning," beyond pure physics. To understand this he needed to find a *"neutral* language," understandable by psychologists as well as physicists, into which to translate the concept of spectral lines. He was particularly interested in his dreams of doublets—where the fine structure appears as two spectral lines. He related this to our experience of the division into two components at the moment of birth when, like the doublet splitting, a child becomes an independent existence. It is also linked to doubling in a psychic sense in which the "new conscious content indicates a mirror image of the unconscious"—the conscious as the mirror of the unconscious.

In 1953 Pauli had a particularly memorable dream about spectral lines. In it, he and Franca were observing an experiment whose results appeared as spectral lines on a photographic plate. One of the lines had a fine structure. He described it thus: the dream "contains a *favorable* indication—namely, the fine structure of the second line." His interpretation was: "What this does is to indicate the beginning of an assimilation of an unconscious content into consciousness." In the dream, he added, "My wife says that she finds this very interesting." In other words, he took the dream to mean that his unconscious was emerging in the conscious. Perhaps by this he meant that his interpretation of the dream was that he was developing some characteristics of Franca's psychological types. Unlike him, Franca was outgoing and in touch with the world.

He noticed that the doublets were like the alternating dark and light stripes on wasps (a great source of fear for him) and tigers. This was, he knew, an archetype. It occurred in Western alchemy and also in India, where he had seen the pattern on Indian temples when he was there with Franca earlier that year. It was an expression of two opposite forces, light and dark, endlessly repeating. In psychological terms it symbolized the tendency of a psychic situation to repeat.

This opposition between light and dark was further clarified by Bohr's complementarity concept, which stated that quantum phenomena could be fathomed in terms of the opposition between complementary pairs—such as wave and particle. Bohr had been sure that complementarity went beyond physics and was basic to all of life, where the complementary pairs of life/death, love/hate, and yin/yang played a key role. All this, said Pauli, "seems to point to a deeper archetypal correspondence of the complementary pairs of opposites." And it was symbolized by the splitting of spectral lines into two, a separation defined by 137. This reinforced his belief that 137 was an archetypal number.

It also reminded him of the patterns of lines that form the basis of the Chinese Book of Changes—the *I Ching*.

I Ching

The *I Ching*, a Chinese oracle, was written four thousand years ago. It was translated into German by Richard Wilhelm, a Sinologist and a close friend of Jung's. Jung considered that it revealed insights into chance occurrences that cannot be understood using the Western concept of causality.

The basic structure of the *I Ching* consists of sixty-four combinations of six broken and unbroken lines, laid out one above the other: the hexagram. The broken line represents yin, the feminine principle, the unbroken one, yang. To consult the oracle, one builds up a hexagram by casting three Chinese coins six times. The inscribed side of the coin counts as yin and has a value of two, the other side as yang with a value of three. One then looks up that hexagram in the *I Ching*. What the oracle has to offer for any one hexagram is extremely gnomic and requires careful interpretation.

The prediction relates to many factors, foremost that the world about us emerges from a struggle of opposites—yin and yang—signifying good/evil, light/darkness, love/hate, man/woman, and other dualities, quite foreign to the rationalism of Western thought. Jung often emphasized that to the Western mind the whole process seems like nonsense. But Western science also has little light to shed on the psyche. Thus other ways of knowing have to be considered. Jung believed that the message of a hexagram—written thousands of years ago—can illuminate the hidden qualities of the present moment, a coincidence in time that cannot be explained by Western physics.

Pauli, too, consulted the *I Ching* for advice "when interpreting dream situations." He noted that to consult the oracle one has to "draw" three times "whereas the result of the draw depends on the divisibility of a quantity by *four*"—those numbers again. Sixty-four, of course, is four multiplied by four three times—4^3. This brought Pauli back to the world clock in which the "motif of the permeation of the 3 and the 4 was the main source of the feeling of harmony."

In his writings Wilhelm had discussed the significance of "magical pictures of trees in rows," relating the image to hexagram 51—"The Arousing" (Shock, Thunder)—in the *I Ching*. In this hexagram the two trigrams—the top and bottom sets of three lines—consist of two broken lines (like two doublets) on top of an unbroken one, which seems to push them violently upward, as if in the awakening of a life force. The text reads:

> The superior man sets his life in order
> And examines himself.

It was a message Pauli was determined to take to heart.

11

Synchronicity

The riddle of the electron

IN ANCIENT TIMES, matter was thought of as being the mother of all things. From this alchemists derived the notion of the prima materia (prime matter), which is uncreated and which therefore contains the attributes of God. In modern physics, conversely, matter has become entirely ephemeral in that it can be created and destroyed, as in the spontaneous creation and annihilation of pairs of antiparticles and particles. One of these antiparticles is the antiparticle of the electron, the positron, which possesses exactly the same properties as the electron except that it has a positive instead of a negative charge. When particle and antiparticle come together they disintegrate in a flash of light. In 1932 the positron— first predicted by Paul Dirac in the famous Dirac equation—had been discovered in the laboratory.

This supported Pauli's view that there was no foundation for a view of life based on the pre-eminence of matter. Einstein symbolized his discovery that mass—that is, matter—and energy were equivalent in the equation $E = mc^2$. Here solid mass is replaced by energy, which has no form. Energy is indestructible and outside of time, and as a result the total quantity of energy always remains the same. This is known as the law of conservation of energy. But one of the astounding results of relativity theory is that there is no law of conservation of mass.

Although energy is timeless, it appears in space and time in particular ways. In quantum physics the energy of a spectral line is proportional to its frequency, that is, the number of oscillations of light per time interval. Imagine you have isolated a single hydrogen atom whose lone electron occupies a stationary state above its ground state or lowest level. The atom is said to be in an excited state. In nature the preferred mode of being is equilibrium. The lone electron will eventually drop to its lowest level and emit light. This can be measured in the laboratory as a spectral line. Observing the atom over a long time results in a very narrow spectral line with a precisely determined energy. Information has been lost, however, because the scientist doesn't know when the electron made its transition to the lowest level. Conversely, observing the atom in its excited state over a short time results in a broad spectral line whose precise energy cannot be determined—there is a spread of energies. But at least now the scientist knows the precise time at which the transition to the lowest level was made.

In other words, the more precisely you know the energy of a spectral line which sparks when an electron jumps from a higher to a lower orbit in an atom, the less precisely you can measure the time that it took to make the transition. There is an uncertainty relationship between energy and time, similar to the one between position and momentum that Heisenberg discovered.

Pauli referred to these two axes—energy and time—as "Indestructible energy and momentum" versus "Definite Spatio-Temporal Process" and saw them as complementary aspects of reality, in that a little of each is always present to a greater or lesser degree.

Pauli's dreams had convinced him that there was a relationship between the frequency of spectral lines, particularly doublets, and the tension between pairs of opposites such as conscious and unconscious. Energy, which is outside of time, is complementary to processes occurring at definite intervals of space and time, and similarly there is a complementarity between the archetypal psyche which exists throughout time (the timeless collective unconscious) and our own individual conscious psyche, or ego, which exists over specific time intervals in our daily life. (He abbreviated the ego as "self-awareness" and the archetypal psyche as "time.")

Pauli laid all this out as a mandala in the shape of a cross. From this he deduced that the laws of physics are a projection onto the psyche (the conscious/unconscious) of an archetypal association of ideas arising from the collective unconscious: in other words, a clash of the four opposing concepts that he depicts at opposite ends of the two crosses.

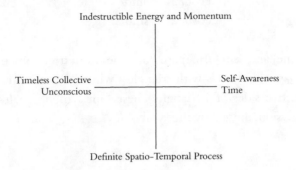

Pauli's preliminary mandala showing the collective unconscious and events in space and time.

The mandala lays out the fundamental complementarity at the heart of both psychology and quantum physics. Bohr had spoken of "the general difficulty in the formation of human ideas, inherent in the distinction between subject and object." In quantum physics, the person making the measurement and the measuring apparatus affect whatever is being measured. Similarly in psychology, the psychologist can never *really* know the unconscious through psychoanalysis. He must always interpret the results of his questions and inevitably he himself will affect his conclusions. Data can never be understood except through the lens of a theory.

In 1948, around the time of the spring equinox, Pauli had two dreams. For him the equinoxes, he said, were times of "relative psychic instability, which can manifest itself both negatively and positively (creatively)." The dreams that arose at those times were always of particular significance.

His dreams were full of mathematical symbols. In one of them i appears, i being the square root of -1: $\sqrt{-1}$. i is an "imaginary number" because it is not one of the numbers we use in daily life—the so-called real numbers. Nevertheless, *i* often functions to unify complicated formulas by making them more compact.

In one of his dreams a woman brings Pauli a bird. It lays an egg that then divides into two eggs. Then he notices that he has another egg in his hand, which makes three. Suddenly the egg in his hand divides into two. He now has four; a quaternity has appeared. Before his eyes the four eggs morph into four mathematical symbols, in two groups, side by side:

$$\cos\delta/2 \quad \sin\delta/2$$
$$\cos\delta/2 \quad \sin\delta/2$$

"cos" (cosine) and "sin" (sine) are quantities from trigonometry (a form of mathematics that deals with triangles) while "δ" (delta) is the angle formed by two sides of a triangle. These four symbols coalesce into a single expression, unified by the symbol *i*:

$$\frac{\cos\delta/2 + i\sin\delta/2}{\cos\delta/2 - i\sin\delta/2}$$

This expression is well known to mathematicians.

In his dream he turns this expression into an equation:

$$\frac{\cos\delta/2 + i\sin\delta/2}{\cos\delta/2 - i\sin\delta/2} = e^{i\delta}$$

where *e*, the "base of natural logarithms," has a numerical value of 2.71828 ... (referred to as an "irrational" number in that the group of numbers "1828" never repeats); and $e^{i\delta}$ has a magnitude of 1. The insertion of *i* into these sets of four has created a unity.

Reflecting on the eggs in his dream, Pauli realized that it was precisely what Maria Prophetissa, the early practitioner of alchemy, had described some seventeen centuries earlier: "One becomes two, two becomes three, and out of the third comes the One as the fourth." This transformation, he noticed, "typically comes about for me through mathematics."

Pauli's interpretation of this whole dream is far removed from mathematics. Describing it to Jung, he explains that $e^{i\delta}$ is a number that always lies on a circle of radius 1. Through the power of the mathematical symbol *i*, a mandala has appeared in the form of a circle. In Pauli's dream *i* "has the irrational function of uniting pairs of opposites"—the cosine and sine functions arranged in two groups of opposites—"and thus producing wholeness." But *e* too is "irrational," it is an irrational number. This shows, he says, that mathematics "is a symbolic description [of nature] par excellence." Mathematical symbols are the perfect way to unite and represent counterintuitive features of the quantum world, such as the wave-particle duality, which can never be visualized.

Reflecting further, Pauli suggests that the successive splittings of the eggs are analogous to the splitting of spectral lines. When one examines the fine structure of a spectral line, the spectroscope shows that what appears to be a single line is actually two and that the spacing between the two lines is defined by the number 137. In that case, could it be, he wondered, that two was the primal number in physics, not four? In both physics and psychology there were complementary opposites suggesting that two was the predominant number in the psyche as well. But four— the quaternity—had appeared in his dreams, signifying the wholeness of the material world and our conscious knowledge of it as well as the unconscious. Pauli's discovery of the fourth quantum number indicated precisely the need for this wholeness and therefore, although it was surprising at first, it should have been expected all along, given that four was the archetype of completeness.

i, the square root of −1, which unifies the various elements in Pauli's dream, also appears in Schrödinger's wave function (the solution to Schrödinger's equation). Schrödinger's wave function depends on *i* and unifies the wave and particle properties of matter as well as being the essential ingredient in making measurements in quantum physics.

This dream reinforced Pauli's conviction that quantum physics ought somehow to form part of a more comprehensive, bigger world picture. It referred only to phenomena that could be described by mathematics and focused its attention on what could be measured in the laboratory, and did not take into account notions such as consciousness. It dealt only with the realm of inanimate matter and thus the anima had been excluded. This excluding of the anima was exactly what Kepler had done when he set about developing modern physics and why Pauli eventually came to side with Fludd.

Parallels and coincidences

Any discussion of dreams, physics, and psychology, Jung believed, required examining the notion of time, and in particular "synchronicity," a concept which he had been exploring since his early fascination with parapsychological phenomena as a medical student.

In the following years Jung had read deeply in mythology and alchemy where he developed the notion of "one world"—the *unus mundus*. If there were one world, he reasoned, surely there should also be one mind, which he identified as the collective unconscious of humankind.

When he consulted the *I Ching*, the advice appeared to be called forth by the moment. If he asked the same question a second time—at a different moment—the advice might be quite different. If the *I Ching*'s answers had any meaning at all, then how did "the connection between the psychic and physical sequence of events come about?" he wondered.

In the 1920s Jung began to look seriously into parallels between out-of-body occurrences and mental states. One notable example occurred in 1928, when Jung drew a mandala that looked to him very Chinese and on the exact same day received in the mail Richard Wilhelm's manuscript of his translation of the *Secret of the Golden Flower*. To Jung, that was what synchronicity was all about. In the Western world, we usually assume that events develop sequentially, one after the other, by a process of cause and effect. But Jung was convinced that as well as a vertical connection, events might also have a horizontal connection—that all the events occurring all over the world at any one moment were linked in a kind of grand network. Thus when one threw the coins to consult the *I Ching*, the throwing of the coins coincided with one's feelings at that precise moment and the answer reflected the truth of that moment.

The turn-of-the-century adventurer John William Dunne reported experiences that were not explicable within the usual sequential framework of time. In his book *An Experiment with Time*, published in 1927, he wrote of recurrent dreams in which he foresaw tragic events such as disastrous military expeditions and volcanic explosions which resulted in large-scale ruin and deaths. In 1902, at the age of twenty-seven, while a soldier in the second Boer War, he had a dream in which he saw the catastrophic volcanic explosion of Mount Pelée on Martinique. When he tried to warn the French authorities, they turned a blind eye. A few days later he read about the disaster in the newspaper. The most extraordinary thing was that the dream had occurred not at the time of the eruption, but several days later when the paper was on its way.

Dunne proposed that time might not always unfold in a straight line, as in physics. Perhaps during sleep the psyche was freed from the rigid march of time and time took on a multidimensionality in which haphazard combinations could occur. Jung wrote glowingly to Pauli about Dunne's clairvoyance.

In May 1930, in a memorial lecture for Richard Wilhelm who had recently died, Jung spoke of the philosophy behind the *I Ching*, which Wilhelm had translated. He said, "The science of the *I Ching* is based not on the causality principle but on one which—hitherto unnamed because not familiar to us—I have tentatively called the *synchronistic* principle."

Synchronicity, he often boasted, was "one of the best ideas" he ever had. His experiences as a psychologist had convinced him that scientific laws of causality were insufficient to explain "certain remarkable manifestations of the unconscious." Jung was well aware, however, that physicists would have nothing to do with acausality. He was eager to find a way to back up his developing ideas with scientific rigor. He desperately needed guidance. It was at that point that he met Pauli.

Synchronicity and telepathy

In 1934 Pauli put his friend and successor at Hamburg, Pascual Jordan, in touch with Jung. Jordan was a highly respected physicist who had carried out ground-breaking research in the new quantum mechanics. He was also a rather eccentric character with a pronounced stutter; his wife used to attend his lectures and make bird noises to distract him whenever he lost control of his words. He modeled his hairstyle on Adolf Hitler's, which, unfortunately, reflected his politics. (This was the friend to whom

Pauli wrote the postcard addressed to PQ – QP.) In the 1930s Jordan moved from pure physics research into studying the effect of quantum physics on biology and also began looking seriously into telepathy.

Pauli sent Jung one of Jordan's articles, which the editor of the highly respected scientific journal *Die Naturwissenschaften* had asked him to referee. It was on the subject of parapsychology. Pauli was skeptical but also curious. He told Jung about Jordan's physics credentials, his speech defect, and his personal problems. Jordan often complained that he had "run out of luck in physics," which was what had led to his "preoccupation with psychic phenomena."

Jung was ecstatic that a physicist of such high repute was interested in the paranormal. Jordan's interpretation of telepathy was that it was sender and receiver sensing the same object simultaneously in a common conscious space. Jung, conversely, considered that the instance of telepathy occurred not in a conscious space but in a common unconscious with only one observer "who looks at an infinite number of objects," not just one.

Jung wrote directly to Jordan in glowing terms, congratulating him on his interest in psychology. He drew his attention to Wilhelm's translation of *The Secret of the Golden Flower*. "Chinese science," he wrote, " is based on the principle of synchronicity, or parallelism in time, which is naturally regarded by us as superstition." Jung also suggested that Jordan look at the *I Ching*.

Synchronicity in physics and psychology

It was not until many years later, in 1948, that Pauli and Jung began to look deeply into synchronicity. In a letter, Pauli asked whether Jung would use the term synchronous, or synchronistic, if there was a gap of time between the dream and the external event it predicted. Jung replied, "nowadays, physicists are the only people who are paying serious attention to such ideas." Pauli suggested Jung record his thoughts on the matter. Jung happily complied and sent Pauli a thick manuscript to read. Four years later it appeared as "Synchronicity: An acausal connecting principle," in a book that Jung and Pauli coauthored: *The Interpretation of Nature and the Psyche*. The book also contained Pauli's essay on Kepler. Before that, however, Jung had to undergo tough criticism from Pauli.

The scientific basis that Jung proposed for synchronicity lay in one of the most dramatic implications of quantum physics: that the coordination in space and time of any atomic process and its causal description are mutually exclusive. One can choose one or the other, but not both. As we saw above, the reason lay in the measurement process itself, in which the measuring apparatus and the "system being measured" (for example, the electron) were inextricably linked. This resulted in unavoidable errors and was at the root of the statistical basis of quantum physics. Moreover, the characteristics of the "system being measured" underwent an unalterable change in such a way that all its individual features were lost. Deriving his knowledge of science from Pauli, Jung interpreted this as showing that there could be other connections of events in space and time besides the causal connection. Perhaps the same applied to the psyche.

Rhine's experiments in ESP

Jung was also intrigued by the experiments that Joseph Banks Rhine, an American psychologist at Duke University, North Carolina, performed in the 1930s and recorded in a book called *Extra Sensory Perception*. It was Rhine who coined the acronym ESP.

Rhine conducted a series of experiments in which a person drew a card from a shuffled deck and a test subject in another room tried to guess what it was. The subjects often achieved astounding results, guessing the correct cards 40 or 50 percent of the time. One subject was 100 percent correct.

Jung examined the archetypal basis for Rhine's experiments. One thing that was striking was that the number of successes decreased sharply after the first attempts and eventually disappeared as the number of tests increased. Rhine attributed this to the subject's lack of interest as time wore on. But when interest and enthusiasm were revived, along with the subject's belief in ESP, the number of successful guesses rose.

Pauli suggested that the decline in the success rate of Rhine's subjects was due to the "pernicious influence of the statistical method," by which he meant that the statistical approach only dealt with large numbers of successful and unsuccessful tests. The size of the sample was so huge that the fact that some subjects had achieved an extraordinarily high success rate simply disappeared in the welter of figures and "the actual influence of the psychic state of the participants" became imperceptible.

Added to this, the mechanical nature of the experiments meant that the participants eventually got bored. As their interest in the experiment decreased, so did their psychic power, thereby blurring the initially exciting valid results. Nevertheless, this was clearly another example of complementarity, in that "any connection between causality and synchronicity can never be ascertained," the two by definition being mutually exclusive. Acausal—that is, synchronous—events were certainly rare, in the realm of single figures. But they existed nonetheless.

Jung and Pauli were impressed by the quality of Rhine's professed scientific standards and marveled at how his data had stood up to criticism. Pauli could see that this was important for Jung's theory of synchronicity, based on the claim that scientific causality was not the complete story. But he could not "see any archetypal basis (or am I wrong there?)," he wrote.

Jung took up the challenge.

Jung's astrology experiment

Around this time, Jung was conducting an astrology experiment. Having gathered data on 180 married couples, he constructed the horoscopes of each partner, hoping to find out whether the dates of birth and positions of the sun and moon for each actually correlated in the way that astrology predicted, within statistical bounds. If they did, then that could provide a scientifically verifiable proof of astrology.

Pauli was uncomfortable with this. He pointed out to his friend, the scientist Markus Fierz, who was assisting Jung with statistical calculations on his astrological data, that Jung had not included the effect of irrational factors entering from Jung's own unconscious and that of his co-workers. "It's a curious thought that it is we physicists who have to call the attention of the psychologists of the unconscious *to this*," he wrote. In the published version of his experiment Jung admitted as much, giving instances where his state of mind and that of his co-workers affected the way in which they constructed the horoscopes. Sadly, his results did not bode well for constructing a scientific basis for astrology. Small samples, however, produced good results, which Jung interpreted as demonstrating the predominance of archetypes from astrology: even though people think they are consciously choosing their partners, in fact they are not. Just as when one consults the *I Ching* and in Rhine's ESP experiments there is no cause-and-effect connection. Jung was the eternal optimist.

In 1949 Pauli wrote to Fierz:

> May this now be a good omen as regards my relationship with physics and psychology, which undoubtedly is among the peculiarities of my intellectual experience. What is decisive to me is that I *dream* about physics as Mr. Jung (and other nonphysicists) *think* about physics. The danger of this situation lies in Mr. Jung publishing nonsense about physics and could moreover quote me in the process. The thing is to prevent this and to turn the matter to advantage. I simply *cannot* evade it! But every time I have talked to Mr. Jung (about the "synchronistic" phenomenon and such), a certain spiritual fertilization takes place.

Pauli was worried that his reputation might suffer if Jung published material on physics that made no sense and that quoted him as confirmation. But their conversations were far too fruitful to dream of abandoning them. Above all he was gripped by the notion of finding a link between quantum physics and psychology—which surely lay in synchronicity.

The scarab and the birds

Wrestling with the concept, Pauli discovered that he found it useful to make a distinction between chance occurrences of synchronicity and occurrences of synchronicity brought about by consulting oracles such as the *I Ching*. For chance occurrences he used the term "meaning-correspondence" rather than "synchronicity," which Jung tended to use synonymously with "simultaneity."

In reply, Jung brought to his attention two examples of synchronism in which he was able to identify "some archetypal symbolism at work . . . which cannot be explained without the hypothesis of the collective unconscious."

The first concerned a woman patient whose animus (that is, her male aspect, the female equivalent of the male anima) clung to a stubbornly logic-based view of reality. She had already been to two analysts before Jung. He was having no success either until one day she told him about a dream of a scarab she had had. At that same moment Jung heard a tapping on the windowpane. He flung open the window and an insect flew in. Jung caught it. It was of the scarab family. To Jung this was not a chance happening but a meaningful coincidence. The patient had been disturbed by the dream scarab and the sudden appearance of a real one completely shattered her stubbornly rational attitude. The scarab bursting in through the window allowed her animus to burst its logical chains and place her on the path to psychic renewal—entirely appropriate, said Jung, given that the scarab is an ancient Egyptian symbol of rebirth. It was an example of a psychic state in the observer coinciding with an external event that corresponded to that psychic state.

The other example of synchronicity concerned the wife of one of Jung's patients. She told Jung that when her mother and grandmother died, on each occasion a flock of birds had gathered outside the window of the room. Some time later, Jung noticed that her husband had symptoms of an impending heart problem and recommended that he see a specialist. The specialist, however, could find no problem. On his way back the man collapsed in the street. Shortly after he had set off to see the specialist a large flock of birds had alighted on the house. His wife immediately recognized this as a sign of her husband's impending death.

Jung noted that in the Babylonian Hades the soul is adorned with feathers and in ancient Egypt the soul was considered to be a bird. It was an example of a psychic state coinciding with a corresponding, not yet existent, future event.

In Rhine's experiments it was the subjects' determination to achieve the impossible—to show that ESP existed—that caused them to tap into their unconscious. Dunne's dreams showed that the psychic state can coincide with an event (like the volcanic eruption) when the subject is asleep. In both cases quieting or closing down the conscious mind enabled the subject or dreamer to open the unconscious to the external world and to allow archetypes to emerge. Divinatory procedures, such as consulting the *I Ching*, required this same mental condition. In each instance of synchronicity that Jung observed, an archetype appeared—the scarabs, the birds. "The effective (numinous) agents in the unconscious are the archetypes. By far the greatest number of spontaneous synchronistic phenomena that I have had occasion to observe and analyze can easily be shown to have a direct connection with an archetype," Jung wrote.

Pauli was still doubtful about Jung's use of the term "synchronistic" to mean "at the same time." Surely this held only for experiences in the first category (an external event coinciding with a psychic state, as in the case of the scarab). While he was mulling over this problem, Pauli had a dream. It was October 1949.

The stranger/Merlin appears

The dream concerns a "stranger" who appeared in earlier dreams as the "blond" man.

In Jungian terms he is the voice of the collective unconscious, the background archetypes given shape—constellated—by twentieth-century scientific concepts, and represents authority. He often comments that modern physics is inadequate and incomplete and is able to move back and forth between the physical and the psychic, the conscious and the unconscious. He is an intermediary, like Hermes in alchemy, a "psychopomp," who moves between the dark and light worlds.

While working on Kepler, Pauli read *Romans de la Table Ronde* (Stories of the Round Table), containing the legends of the Holy Grail. He was struck by the similarity between the "stranger" and the wizard, Merlin. Emma, Jung's wife, was also interested in the Grail. Pauli wrote to her:

[The stranger] is a spiritual light figure with superior knowledge, and on the other hand, he is a chthonic [dark] natural spirit. But his knowledge repeatedly takes him back to nature, and his chthonic origins are also the source of his knowledge, so that ultimately both aspects turn out to be facets of the same "personality." He is the one who prepares the way for the quaternity, which is always pursuing him. . . . He is not an "Antichrist," but in a certain sense an "Antiscientist," "science" here meaning especially the scientific approach, particularly as it is taught in universities today. . . . My branch of science, physics, has become somewhat bogged down. The same thing can be said in a different way: When rational methods in science reach a dead end, a new lease on life is given to those contents that were pushed out of time consciousness in the 17th century and sank into the unconscious. [The stranger] happily uses the terminology of modern science (radioactivity, spin) and mathematics (prime numbers) but does so in an unconventional manner. Inasmuch as he ultimately wishes to be understood but has yet to find his place in our contemporary culture, he is, like Merlin, in need of redemption.

In some ways the "stranger" seems to represent Pauli himself—not surprisingly, for he springs from the collective unconscious, which, according to Pauli, has now been given "a new lease on life."

Of the stranger, he wrote to Aniela Jaffé, Jung's secretary:

Like Merlin, he *knows* the future, but cannot change it. . . . In my opinion, however, *man can alter* the "future." . . . I want to recognize [Merlin], talk to him again, bring his redemption a little nearer. That, I believe, is the myth of my life.

For Pauli rational methods had reached a dead end and were no longer the tools that would enable him to change the world. Rather, the magical world of Merlin with its search for the quaternity held the key. If Pauli could only come face to face with him, he could "bring his redemption a little nearer" and so, too, with the "stranger" who could not speak a language that could be understood by everyone. Pauli believed that to move forward in examining the human psyche he needed to fuse physics with psychology. This was the "myth of [his] life," no less heroic than that of Merlin.

In Pauli's dream, an airplane lands and some foreigners step out, among them the stranger. He tells Pauli, "You should not exaggerate your difficulties with the notion of time. The dark girl has only to make a short journey, in order to determine the time!"

Jung's interpretation of this dream was that the airplane represented Pauli's intuition and the foreigners his "not-yet-assimilated thoughts."

The dark girl is Pauli's anima. She has to "make a short journey," that is, change her place in order to achieve definite time. At present "she has no definite time," meaning that she lives in the unconscious. She has to transplant herself "into consciousness in order to be able to define time." The stranger wants Pauli's anima—the feminine side of his personality— to study the mathematics of whole numbers which are the *"archetypes of order,"* in order to understand synchronicity. In this way, Pauli will be able to move toward a unification of physics and psychology, the reverse of Kepler's materialistic worldview so deplored by Fludd.

Jung concluded his letter with a new quaternary diagram:

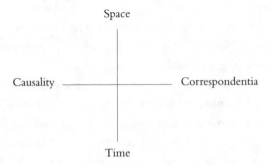

Jung's response to Pauli's mandala.

In this, Jung takes space and time as complementary. Opposite the causality of physics he places "correspondentia"—the correspondence between the psychological and the physical view of life, including synchronicity.

Back to Bohr's complementarity principle

Bohr, too, in his view of complementarity had something to say about causality:

> The very nature of the quantum theory ... forces us to regard the space-time coordination and the claim of causality, the union of which characterizes the classical physical theories, as complementary but exclusive features of the description, symbolizing the idealization of observation and definition respectively.

Classical physics combines how a system develops in space and time with causality (meaning a logical chain of cause and effect). The mathematical structure of Newton's laws of motion permitted the path of an object to be traced in space and time with, in principle, perfect accuracy, that is, to predict the paths of cannonballs, falling objects, and planets. This is the law of causality. To use it the scientist needs only two pieces of information: where the object was and how fast it was moving when the process began. Knowing that a stone was six feet off the ground and dropped from a resting position, we can predict where it will be as it is falling and when it will hit the ground.

Yet Heisenberg's uncertainty principle asserts that it is impossible to make exact measurements of an electron's position and its momentum in the same experiment. Thus according to quantum theory it is an impossibility—an idealization, as Bohr puts it—to combine a description in space and time with causality.

According to Bohr's complementarity principle, the description in space and time of a physical system (such as a quantum of light hitting an electron in the same way that two billiard balls strike each other) and causality (predicting where the electron and light quantum will be after they bounce off each other) are complementary and mutually exclusive. But every scientific theory must be causal or else it cannot make predictions, which are essential to science.

So can there be predictability, that is, causality, in quantum mechanics? The conservation laws of energy and momentum state that the amount of energy and momentum in a system cannot change. Scientists can apply these laws to predict the final condition of a system from its initial state.

If a quantum of light striking an electron is like two billiard balls, then it should be possible to use the laws of conservation of energy and momentum to work out where to set up instruments to detect the light quantum and the electron after they collide. In quantum physics the law of causality of classical physics—which requires precise measurements of position and momentum in the same experiment—is replaced by predictions made by the laws of conservation of energy and momentum.

A new mandala

In response to Jung's analysis of his dream, Pauli commented that he agreed that the stranger conveyed a holistic view of nature quite different from the "conventional scientific point of view." Unlike his colleagues, Pauli wrote, he considered the quantum mechanics as incomplete. What was required was a fusion with psychology. He had "no shortage of 'not-yet-assimilated thoughts'," he added wryly.

He disagreed, however, with Jung's mandala primarily because it showed space and time as separate, whereas scientists understood that they were one—the space-time continuum. He suggested another one which included space-time while retaining the psychological element of Jung's:

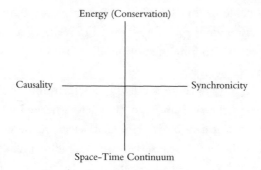

Pauli's suggested improvement to Jung's mandala.

Here he lays out complementary pairs, causality—the chain of cause and effect—against synchronicity; and conservation of energy against the space-time continuum, in agreement with Bohr's complementarity principle. Classical physics pairs causality with a description in space and time. But this is an idealization. And so Pauli set in its place the law of conservation of energy; to be more precise the law of conservation of momentum should be included too.

Synchronicity in physics and psychology

The essential question Pauli felt needed to be asked was, "How do the facts that make up modern quantum physics relate to those of other phenomena explained by [Jung] with the aid of the new principle of synchronicity?" How did quantum physics sit in relation to synchronicity and other psychological phenomena. Both types of phenomena, he noted, went beyond "classical determinism."

In Pauli's mandala, energy and space-time, and causality and synchronicity, are complementary but mutually exclusive, like light and dark and life and death. Both arms are necessary. It is the tension between them that gives physical meaning to reality.

Pauli also noted that when Jung used "physical terms to explain psychological terms or findings," to Jung these were "dreamlike images of the imagination." Jung, for example, referred to radioactivity as a physical analogy for a coincidence in time—total nonsense to a physicist. Pauli proceeded to explain to Jung the notion of probability in quantum physics using radioactive decay.

In quantum physics there is a law for determining how many of a large sample of nuclei will undergo radioactive decay by emitting particles and light. But it cannot determine at what precise point in time a single nucleus will decay because it is impossible to investigate a single atom and how it develops in space and time. In other words, individual events are outside of the chain of cause and effect.

On average, half the total sample will decay in the "half-life"—a period of time that is a characteristic property of each radioactive element. After another half-life, another half of the sample will decay. But it is impossible to know when any particular nucleus will decay. To find out, one has to carry out a measurement on the system that causes decay rather than measuring when the decay naturally occurs. The law of radioactive decay is built up out of the probability of each nucleus decaying, that is, it is statistical. Moreover, the statistical regularity—the prediction of when half the sample will decay—is reproducible and has nothing to do with the psychic state of the experimenter. This is the exact reverse of experiments (such as Rhine's) on synchronicity, which turned up a small number of examples of synchronicity that when viewed statistically were so few as to be negligible. The regularity of the half-life period could be ascertained only when there was a large number of cases, whereas in the Rhine experiments synchronicity appeared only in a small number.

Pauli's explanation of probability in radioactive decay was also a reply to a query Jung had raised: what light does synchronicity throw on the "half-life phenomenon of radium decay?" Just as it was impossible to tell whether any one radium nucleus had decayed, similarly it was impossible to identify the precise connection of one individual with the collective unconscious. The moment when an individual nucleus decays is not determined by any laws of nature and exists independently of any experiments. Nevertheless, when someone carries out the experiment this moment becomes a part of the experimenter's time system. The very act of measuring whether an individual nucleus has decayed alters its condition and perhaps even causes it to decay.

Pauli suggested that the state of the individual radium nucleus before the experiment was carried out might correspond to the relationship of an individual to the collective unconscious through archetypal content of which the individual was unaware. As soon as one tried to examine an individual consciousness, the synchronistic phenomenon would immediately vanish.

Pauli's understanding of synchronicity firmly separated it from processes in physics. Jung offered quite a different definition: perhaps "synchronicity could be understood as an *ordering* system by means of which 'similar' things coincide, without there being any apparent 'cause'. . . . I see no reason why synchronicity should always just be a coincidence of two psychic states or a psychic state and a nonpsychic state." In opposition to Pauli, Jung suggested broadening the concept of synchronicity to include every sort of coincidence, whether between two psychic states or two elementary particles. He was intrigued by the fact that it is impossible to predict when an individual nucleus will decay, which opens up the possibility of phenomena in individual atoms that are beyond cause and effect.

Jung pointed out that modern physics had shown that the connection between space and time was crucial. In our daily world of consciousness, space and time remain two separate entities. "No schoolboy would ever say that a lesson lasts for 10 km," wrote Jung. The world of classical physics had not ceased to exist—we still use Newtonian science to build bridges, for example. Similarly, despite Jung's and Freud's discovery of the unconscious, "the world of consciousness has not lost its validity against the unconscious." Our commonsense perceptions about the world—of space and time as separate and consciousness as our preeminent experience—were still valid.

To replace the mandala he had drawn showing the world of consciousness which experiences space and time as separate, Jung proposed a more complex one that he devised with Pauli's help.

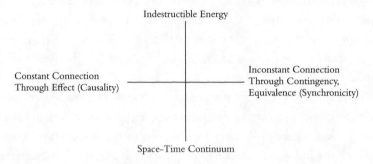

Jung's mandala covering all instances of synchronicity.

This, claimed Jung, satisfied the "requirements of modern physics on the one hand and the psychology of the unconscious on the other hand."

Jung's definition of synchronicity—that is, "inconstant connection through contingency, equivalence (synchronicity)"—Pauli replied, seemed to cover every system that was beyond cause and effect, including quantum physics. Pauli was intrigued because Jung's broadened definition of the archetype seemed to offer a means to develop a unified view of the world. Did this mean that the concept of the archetype, too, could somehow be applied to quantum physics? Perhaps the "archetypal element in quantum physics [was] to be found in the mathematical concept of probability."

Jung enthusiastically agreed that mathematical probability must correspond to an archetype. Bringing archetypes and synchronicity together, he suggested that the archetype "represents nothing else but the probability of psychic events." Although all of us are born with a collective unconscious made up of archetypes, it is not inevitable that any single archetypal image will actually appear in our consciousness. It is only highly probable—not inevitable—that patients recovering from deep depression will draw mandalas.

The law of probability in quantum physics is a law of nature and laws of nature contain the patterns of behavior of the cosmos. Given that the archetype is also a pattern of behavior, does this mean that laws of nature have their bases in psychic premises? And how do archetypes enter our human minds in the first place? Jung suggested that they were "out there," ready to be plucked out of the air, and in this way entered our minds. We are all, after all, merely small elements in one world. The origin of the word is immaterial, Jung insisted; it's what the archetypes can do that is important.

Returning to the ever-fascinating issue of threes and fours, Jung perceived that quantum physics widened the threesome of classical physics—space, time, and causality—to include synchronicity, thereby becoming a foursome. This happy development solved the age-old problem of alchemists, encapsulated in the "so-called axiom of Maria Prophetissa: Out of the Third comes the One as the Fourth. . . . This cryptic observation confirms what I said above, that in principle new points of view are not as a rule discovered in territory that is already well known, but in out-of-the-way places that may even be avoided because of their bad name."

Jung was delighted to have this unique opportunity "to discuss these questions of principle with a professional physicist who could at the same time appreciate the psychological arguments."

Pauli's Jungian take on Kepler and Fludd

Pauli finally published his essay on "The Influence of Archetypal Ideas on the Scientific Theories of Kepler" in 1952 in a book entitled *The Interpretation of Nature and the Psyche*, which also contained Jung's essay on synchronicity. For Pauli it was a bringing together of all his work—his lectures on Kepler and Fludd, his dreams and conversations with Jung, and his correspondence with Fierz—giving shape to a subject he had been thinking about for twenty-five years.

Pauli's focus was the process of scientific creativity and particularly its irrational side. Though scientific theories are expressed in mathematical terms, the initial discovery of the theory is essentially an irrational—not a rational—process. What role, he wondered, had prescientific thought played in the discovery of scientific concepts and what was the link between the two? He examined the rise of modern science beginning with Kepler, and applying the insights of Jung's psychology. He argued that the process of bringing new knowledge into consciousness involved a matching up of "inner images pre-existent in the human psyche" (archetypes) with external objects. Alchemy had a critical role to play in this process. In Jung's psychology, alchemy offered a way to resolve the tension between opposites. It emphasized the number four (the quaternity) and it also focused on the need to bring about symmetry between matter and psyche.

As Pauli put it, "intuition and the direction of attention play a considerable role in the development of concepts and ideas, generally transcending mere experience, that are necessary for the erection of a system of natural laws (that is, a scientific theory)." This leads him to ask, "What is the nature of the bridge between the sense perceptions and the concepts?" Pauli adds, "All logical thinkers have arrived at the conclusion that pure logic is fundamentally incapable of constructing such a link." At this point Pauli introduces the "postulate of a cosmic order independent of our choice and distinct from the world of phenomena."

He concludes that in the unconscious the place of concepts "is taken by images with strong emotional content"—that is, images of archetypes. Thus, the links between sense perceptions and concepts are archetypes— a word used in a similar sense by both Kepler and Jung. One of the forces driving a person to allow these ideas to bubble up from the collective unconscious is the "happiness that man feels in understanding" nature. Thus Kepler's exuberance over Copernicus's discovery of the sun-centered universe with its mandala-like quality. And thus Pauli also brings in the irrational, or nonlogical, element in scientific creativity, which he had sought for so passionately.

To put it in Jungian terms, Kepler understood the relation of the earth to the sun as being equivalent to the ego and Self. The ego is in psychological terms the center of gravity of the conscious with all its imperfections, while the Self, the totality of the conscious and unconscious, is superior to the ego and associated with archetypal images such as the mandala. No wonder, Pauli commented, that the "heliocentric theory received, in the mind of its adherents, an injection of strongly emotional content stemming from the unconscious." Just as the mind gropes toward a state in which conscious and unconscious are balanced so, too, science gradually becomes more balanced between logic and feeling.

But full centering and the achievement of the Self can occur only when the mandala can rotate. As we saw in Chapter 5, Kepler's mandala lacked the fourth element and therefore could not.

In psychological terms, Fludd offered a more complete view of nature based on the number four, which enabled him to see the world as more than simply a mechanical system governed by mathematics, as Kepler did. Pauli, an astute historian, was well aware of how difficult it would be to put oneself into the mind of Kepler or Fludd, living, as they did, in times radically different from our own. Jung's work offered a way to understand them as different personality types, "a differentiation that can be traced throughout history," wrote Pauli. Kepler was a thinking type, who focused on the parts rather than the whole, while Fludd was a feeling type who sought "a greater *completeness of experience.*" This meant including emotions and the "inner experience of the 'observer'," which Fludd did by taking into account the "power of this number"—namely four.

In the end, however, Fludd was on the wrong path. It was inevitable that modern science would develop as it did, in a way that did not bring about the fully rounded psyche. As Pauli wrote: "In my own view it is only a *narrow* passage of truth (no matter whether scientific or other truth) that passes between the Scylla of a blue fog of mysticism and the Charybdis of a sterile rationalism. This will always be full of pitfalls and one can fall down on both sides."

Certainly, modern scientists could not possibly revert to the archaic and naive view of nature held by Fludd. Yet the current rationalistic view was also too narrow. The only way to broaden it would be "a flight from the merely rational." Science is a product of Western thought. To achieve full understanding of the world about us, it requires an equal input of Eastern mysticism. It is necessary to bring together "the irrational-critical, which seeks to understand, and . . . the mystic-irrational, which looks for the redeeming experience of oneness." These two forms of knowledge represent the struggle between opposites, which is at the basis of alchemy.

"Modern science," wrote Pauli, "has brought us closer to this goal [with] the concept of complementarity," a notion that went beyond the confines of a theory steeped in rational thought. Complementarity offered a view of irrationality and rationality as complementary aspects of the unity of thought.

Ultimately Pauli disagreed with physicists who considered quantum theory as the most complete and final description of nature. It was certainly complete, Pauli agreed, but only within a very narrow domain, with nothing to say about consciousness or life itself. It is ironic, he wrote, that although we have a highly developed and sophisticated mathematical apparatus to understand the world of physics, "we no longer have a total scientific picture of the world." For the deep meaning of quantum physics is that—by definition—"it is impossible ever fully to understand the totality of nature." As Heisenberg's uncertainty principle makes plain, as soon as one grasps one truth—for example, the location of an electron—another truth instantaneously slips from one's grasp—in this case, how fast it is traveling.

"It would be most satisfactory of all if physics and psyche could be seen as complementary aspects of the same reality," he wrote. "To us, unlike Kepler and Fludd, the only acceptable point of view appears to be one that recognizes *both* sides of reality—the quantitative and qualitative, the physical and the psychical—as compatible with each other, and [one that] can embrace them simultaneously."

As CHARLES A. MEIER, the first director of the Jung Institute and editor of the Pauli/Jung letters, recalled, "neither Pauli nor Jung needed much persuading to have their works published jointly," though there had in fact been some pressure on Pauli not to do so. As Pauli wrote to Fierz in 1954:

> Many physicists and historians have of course advised me to break the connection between my Kepler essay and C. G. Jung. . . . I am

indifferent to the astral cult of Jung's circle, but that, i.e., this dream symbolism, *makes an impact! The book itself is a fateful "synchronicity"* and must *remain* one. I am sure that defiance would have unhappy consequences as far as I am concerned. Dixi et salvavi animam meam! [I spoke and thus saved my soul].

Looking back on Pauli's relationship with Jung from a twenty-first-century viewpoint, it is important to remember that Jung, Pauli, and their contemporaries considered Jung's research to be quite as important as Pauli's work in physics. Jung's exploration of the human psyche was just as serious as quantum mechanics' exploration of the physical world. Whereas today we take for granted the conclusions of quantum mechanics, most of us are less ready to accept concepts like synchronicity or archetypes. They are not part of our current currency of belief. But when Pauli and Jung were having their conversations, Pauli took for granted that Jung's research was every bit as weighty and significant as his.

12

Dreams of Primal Numbers

A system of morals for a world without God

PAULI WAS a frequent dinner guest at Jung's. It was a great honor; Jung did not often entertain. He detested small talk and chose his dining companions with care. Similarly, Pauli often refused dinner invitations.

For Jung the dining room and library were the center of the house. The dining room, on the ground floor, was the largest room. In the center, dominating the room, was a large wooden table and at the far end a fireplace that in winter held a roaring fire. Jung had a passion for food, insisting that his meals be exquisitely prepared with only the finest ingredients. After a hearty meal the two friends would sit gazing out at the lawns sweeping down to Lake Zürich, swathed in evening mists, sip an excellent French wine—preferably Bordeaux—and smoke their pipes. Among much else, their talk turned to what seemed to many in those days the growing threat of nuclear war. In the post–World War II years the Cold War was in full swing, and with the availability of nuclear weapons, Armageddon seemed a real possibility.

In 1951, Jung published *Aion: Researches into the Phenomenology of the*

Self. It is a study of archetypal images, especially those of wholeness and quaternity, and looks into Christian symbolism, Jesus Christ, and the problem of evil—a problem that, Pauli wrote to Jung, "has once again become an urgent necessity for modern man."

Commenting on Jung's new book led Pauli into a discussion of religion, philosophy, and the meaning of life. Pauli was well-read, particularly in Schopenhauer and Lao-tse on the philosophical side. Indian and Chinese philosophy were much read in Pauli's circle and Bohr often mentioned Lao-tse, while Eastern philosophy formed part of the nineteenth-century German philosopher Arthur Schopenhauer's view that Pauli admired. But it was Jung who had really sparked his interest in Chinese philosophy and in mysticism. In Pauli's view Bohr's complementarity principle—that the world of elementary particles can be understood in terms of apparently opposing entities, such as waves and particles, actually complementing each other—had been discovered centuries earlier by mystics who believed that reality—though it cannot be seen—can be experienced through the meeting of opposing phenomena. Both the Buddha and Lao-tse taught that one could have a mystical experience without the need for any belief in God. Lao-tse's invisible reality—the Tao (the Way)—possesses neither good nor evil. Pauli approved this lack of duality, which he deemed very un-Western.

He added, "I must confess that specifically Christian religiousness—especially its concept of God—has always left me emotionally and intellectually out on a limb. (I have *no emotional resistance* to the idea of an unpredictable tyrant such as Yahweh, but the excessive arbitrariness in the cosmos implied in this idea strikes me as an untenable anthropomorphism.)" To Pauli it was distasteful to attribute human qualities such as consciousness to God or to postulate a fundamentally evil nature in human beings: "I have a Jewish heritage of psychic capabilities, together with a Catholic sense of ritual and ceremony, together with a definite opinion, *that the entire ideology of Judeo-Christian monotheism is of no use to me,*" he wrote sternly to Jung's assistant Aniela Jaffé.

Pauli was attracted by Schopenhauer's inclusion of Eastern religion, particularly Buddhism, in his writings, especially in his meditations on suffering and desire. In *Aion* Jung speaks of the wheel as symbolizing the cycle of life, an idea "akin to Buddhism." He criticizes the Christian notion of *privatio boni*—of evil as the absence of good. Pauli agreed with this, punningly describing *privatio boni* as "the hole theory of evil" (alluding to Dirac's early view that antielectrons [positrons] are holes in a sea of negative-energy states). While *privatio boni* might be acceptable in Catholicism, Jung believed that analytical psychologists had to "take evil rather more substantially. . . ." He points out that "the Christ symbol lacks wholeness in the modern psychological sense . . . since it excludes the power of evil."

Both Jung and Pauli steadfastly disagreed with people who rejected God only to replace the concept with another name. Thus Schopenhauer replaced God by the unconscious Will, while Hegel employed "intellectual juggling" to raise the issue to the level of philosophical criticism, thus opening the arena of discussion to a myriad of ideas framed in "the megalomaniac language of schizophrenics." All this, wrote Jung, led to the hubris of Nietzsche's superman and "to the catastrophe that bears the name of Germany."

" 'The world as will and representation' means nothing else to me than the world as complementary pairs of opposites," Pauli wrote to Marie-Louise von Franz. Von Franz was a close associate of Jung's who also sometimes worked as Pauli's analyst and became his close friend. What Pauli was looking for was a basis for a system of morals that transcended belief in any deity. He looked to Schopenhauer and Jung for how to proceed. Schopenhauer believed that, at a deep level, all individuals were identical—a precursor of Jung's notion of the collective unconscious. If that were the case surely there could be a theory of ethics and morals that cut across cultures. For both Pauli and Jung this topic was more than academic. It was a matter of urgent concern, fired by the terrible war crimes that had been and were being committed against humanity as well as their horror of the atomic bomb.

Answer to Job

In 1952, a year after *Aion*, Jung published *Answer to Job*. That same year, he published his article, "Synchronicity: An acausal connecting principle," in *The Interpretation of Nature and the Psyche*, the book he coauthored with Pauli. *Answer to Job* is a very personal book in which Jung speaks of the emotions aroused in him by the "unvarnished spectacle of divine savagery and ruthlessness" inflicted by Yahweh on Job. He expands this to include the savagery and ruthlessness in us all and reminds his readers that today, more than ever, the four horsemen of the apocalypse are waiting—in the form of the atomic bomb.

Pauli read the first twelve chapters of *Answer to Job* in one night, September 19, which, as it happens, was close to the equinox. He enjoyed the book; it seemed like light reading. But that night he had a very intense dream.

In his dream he is searching for "the dark girl"—his anima—who for him has always "been the counterpole to Protestantism—the men's religion that has no metaphysical representation of woman." The tension between Catholicism and Protestantism often tormented him in his dreams. It seemed to be a conflict between opposites, one of which (Catholicism) rejects the rational, the other (Protestantism) the anima— the same pairing as Fludd and Kepler, psychology and physics, intuitive feeling and scientific thinking, and Mysticism and Science.

Pauli then dreams of a Chinese woman whom he has seen before and whom Jung interprets as the holistic aspect of the dark girl. She leads him into an auditorium in which "the strangers" await him and gestures to Pauli to go to the rostrum, where he is to give a lecture. As he is mounting the rostrum he wakes up.

Pauli had had similar dreams of a Chinese woman in which he was offered *"a new professorship."* He interpreted the fact that he had not yet accepted the position as indicating that in his conscious mind he resisted it; but his unconscious meanwhile rebuked him for keeping "something specific from the public." He believed strongly that the tradition of science must be adhered to and the rest of his life kept a private matter. With very few exceptions, he never talked about the conversations and exchanges he had with Jung. The "strangers" in the lecture hall in his dream seemed to expect him to speak not only about science, but also about psychology and even ethical problems.

Marie-Louise Von Franz was Pauli's analyst at the time. Many years later she revealed that "Pauli was afraid of the content of his dreams. It frightened him to draw conclusions from what his dreams said. They said for instance that he should stand up for Jungian psychology in public. And that he feared like hell, which I understand. He moved in the higher circles in physics. His colleagues were very mocking and cynical and also jealous of him. If he had stood up for dreams and irrational things, there would have been a hellish laughter and he hadn't the guts to face it. So that was really tragic."

In *Answer to Job,* Jung concludes by saying that man is the focal point around which both science and life revolve. Pauli's comment on this is that the dualities of good and evil, spirit and matter, are all within man. The archetype of the wholeness of man—depicted with the symbol of fourness, the quaternity—is the emotional dynamic that drives all of science. "In keeping with this, the modern scientist—unlike those in Plato's day—sees the rational as both good and evil. For physics has tapped completely new sources of energy of hitherto unsuspected proportions that can be exploited for both good and evil." He is referring, of course, to Einstein's theory of relativity and quantum physics which produced, among much else, the atomic bomb.

Jung respected Pauli's uncomfortable position: "It means a lot to me to see how our points of view are getting closer, for if you feel isolated from your contemporaries when grappling with the unconscious, it is also the same with me." He congratulated Pauli on the effort he put into thinking about analytic psychology "which would give you quite a lot to tell the *strangers* about."

Numbers as archetypes

Pauli imagined that being a physicist by day, his psyche would compensate by throwing up images from psychology by night. But to his surprise, his dreams were full of symbols from physics. He noticed that his dreams contained concepts from Kepler's time. Strangely they "did not simply refer to modern, traditional physics but [represented] a sort of *correspondentia* between psychological and physical fact." Perhaps this was the way to extend terms from physics and mathematics into psychology.

By the early 1950s Jung agreed with Pauli that numbers undoubtedly were archetypes and added that they could "amplify themselves immediately and freely through *mythological statements*," such as the one attributed to Maria Prophetissa. The common ground between physics and psychology was not to do with parallel concepts "but rather in that ancient spiritual 'dynamis' of numbers. . . . *The archetypal numinosity of number* expresses itself on the one hand in Pythagorean, Gnostic, and Kabbalistic (Gematria!) speculation, and the other hand in the arithmetical method of the mantic [divinatory] procedures in the *I Ching*, in geomancy and horoscopy." This Jung wrote to Pauli in 1955.

Mathematicians might argue over whether numbers were originally invented or discovered, just as psychologists debate whether archetypes are innate or acquired. "In my view both are true," wrote Jung. Jung was interested not in what mathematicians did with numbers, "but what number itself does when given the opportunity. This is certainly the method that has proved particularly successful in the field of archetypal ideas." He was curious, in other words, about whether numbers have mystical powers and what these might be. It was certainly a fresh approach to numbers, evidence of the fruitfulness of the collaboration between the two men.

Pauli also discussed his thoughts on psychology with von Franz. She had helped Pauli with translations from Latin to German for his article on Kepler and Fludd. Part of her work with Jung concerned the dreams of the French philosopher René Descartes. She had written an article on the subject and hoped to publish it in Jung and Pauli's joint work, *The Interpretation of Nature and the Psyche*, in 1952. Pauli spoke at some length with her about the article but in the end it was not included in their book. Von Franz was very disappointed and there was a brief disruption in their relationship. But by the end of 1952 they had resumed their friendship, taking long walks, excursions, and boat trips together on Lake Zürich.

Their relationship was a tumultuous one. The two had very different points of view and regularly argued about Pauli's dreams and Jung's psychology. Pauli's analysis was that they were both thinking types. From their correspondence, it seems clear there was a mutual attraction, though Pauli tried to keep his feelings focused on intellectual matters. Some people have suggested that at some point they had a sexual relationship, but von Franz insists otherwise. As to what really happened, we will never know, for Pauli's widow burned all von Franz's letters to him when she discovered them in a box in Pauli's office at the ETH.

Once—it was in 1952—von Franz pointed out to Pauli that "nothing much has been done on the *archetypal meaning of numbers.*" Inspired by her remark, Pauli turned to a book on the history of mathematics, *Science Awakening*, by a Dutch colleague of his, B. L. van der Waerden. There he learned that Pythagorean number mysticism was a *"further development of Babylonian number mysticism."* According to the book, the Chinese originated the idea that even numbers were feminine and odd numbers masculine. The Babylonians took the idea from them. But Pauli considered this line of development "improbable." It was more likely that the notion arose from the "presence of pre-existing (archetypal) images, which are released through numbers." Thus "the archetype of 'oneness' and the archetypes of *opposing pairs*" might lead directly "to the concepts 'even' and 'odd,' " he wrote. Four being the quaternity denotes a wholeness because it includes the anima.

From these archetypes—of oneness and of opposing pairs—Pythagoreans studied the divisions between even and odd numbers and geometrized certain combinations in the tetraktys, the mysterious equilateral triangle representing the equation $1 + 2 + 3 + 4 = 10$.

From whole numbers "emerged such exact abstract concepts" as friendly numbers. In mathematical parlance, a pair of numbers is "friendly" if each of them is the sum of the other's divisors. He told von Franz a story about "friendly numbers": "Someone asked Pythagoras whether he had a friend. He replied *I have two.* He named the friendly numbers 284 and 220." The numbers 284 and 220 are friendly because the numbers by which 220 can be divided to yield a whole number (1, 2, 4, 5, 10, 11, 20, 22, 44, 55, and 110) add up to 284; and the divisors of 284 (1, 2, 4, 71 and 142) add up to 220. Friendly numbers may be just another fascinating piece of mathematics, or they may have some use. Presently no one knows.

"From a psychological point of view what does this 'I have two' mean to you?" asked Pauli. Plenty, he went on, answering his own rhetorical question, because it is "by no means an easy task to find every pair of 'friendly numbers.' " (A few hundred friendly numbers were known in the 1950s; with the help of high-speed computers, twelve million have now been found.) Perhaps a clue is that "here is projected a *psychological* problem connected with numbers." Pythagoras's discovery of this pair was rather extraordinary. Apparently it was an inspired mental leap after a great deal of hard work.

The pair of friendly numbers 284 and 220 were well known. In the Middle Ages talismans inscribed with them were worn by a couple to advertise their love for each other. In Genesis Jacob gave 220 goats to Esau on the grounds that one-half of a friendly pair expressed Jacob's love for Esau. Arab numerologists have written about the practice of carving 220 on one fruit and 284 on another, eating one and then offering the other to a lover as a sort of mathematical aphrodisiac.

Pauli applied this line of reasoning to the dreams about the "strangers," the lecture, and the Chinese woman, which was still nagging at him: "From my earlier dreams it is to be expected that my unconscious soon will be activated, as soon as I am 'cranked up' by means of a suitable lecture." Perhaps thinking about numbers would activate the proper archetypes in his unconscious which would, in turn, enable him to find all the friendly numbers.

The Piano Lesson

In October 1953 Pauli wrote an "Active Fantasy about the Unconscious," which he entitled *The Piano Lesson*. He dedicated it to von Franz. It could be regarded as stream of consciousness, a sort of automatic writing.

He began poetically: "It was a foggy day and for a long time I had been seriously troubled." In his daydream, Pauli is worried about how to bring together physics and psychology. He searches for a neutral language because both physicists and psychologists need to understand not only words, but their meanings. He visits the house of a friend—von Franz. As he enters, a voice shouts, "Time reversal." Suddenly Pauli is back in his home in Vienna, in 1913. There is a piano there and a woman—whom he takes to be his anima—is about to give him a lesson. (The woman is actually his maternal grandmother, of whom Pauli had many fond memories.)

He plays the chord C–E–G and he and the woman discuss four variations of it—all on white keys, all on black keys, a combination of the two, or in a major key. When Pauli asks about the opposition of the white and black keys and the many combinations, the teacher replies, "One can play in minor on the white keys and in major on the blacks, it is only a question of knowing how to play."

In the dream which Pauli had described to Jung, he had been about to give a lecture before an audience of strangers—his "unassimilated thoughts"—who wanted him to speak about psychology. In *The Piano Lesson*, Pauli actually gives the lecture, holding forth on psychology, physics, and biology.

The audience of strangers is eager to learn more, but Pauli leaves and finds himself back with the piano teacher. He tells her that while giving the lecture he has produced a "child" for her. The "child" is a product of his unconscious indicating that the woman is indeed his anima, and represents a new holistic attitude—a fusion of psychology, physics, and biology. The problem is how to make it acceptable to everyone. The answer, Pauli realizes, lies in numbers. He suddenly notices that the teacher has Chinese eyes, like the Chinese woman who, in his earlier dream, had led him to the lecture hall.

A voice says, "Wait, transformation of the evolutionary center." Suddenly the teacher has a ring on her finger marked with the mathematical symbol i—the square root of -1. She tells him that it symbolizes the oneness of the rational and the irrational. But Pauli sees another oneness: i is a key element in Schrödinger's wave function in quantum physics. It describes the wave and the particle properties of matter and symbolizes the unity of wave and particle.

Pauli puts on his coat and hat. He is about to leave when he hears the teacher play a four-note chord. The three-note chord he played at the beginning of the fantasy has mutated into four. A wholeness has been achieved, a child has been produced, and Pauli's thoughts are now focused on the archetype of number, particularly on three and four and how the shift from one to the other brings about a unity.

The two-thousand-year-old problem

Nevertheless, the great problem remained unresolved. As Jung put it, "Oddly enough, the problem is still the same 2,000-year-old one: How does one get from Three to Four?" Pauli's own thoughts on this date back to his discovery of the exclusion principle and the "difficult transition from three to four." That was really the *main work*," he added. Now it began to lead him into other aspects of numerology.

Writing to von Franz, Pauli continued the introspective and numerical mode of *The Piano Lesson*: "*One can not at all proceed from three to four.*" As far as the 2,000-year-old question was concerned, he was stymied. But perhaps he could try a different approach. "*However, one can proceed in various ways to three and four,*" he went on. Then he had a brain wave. Four times three equals twelve; but twelve can also be expressed as two times six. He played with this conundrum at the start of his letter:

A. To the sign of 6.
B. $2 \times 6 = 12$
The professor who shall calculate numerically.

Twelve, he remembered was the number of constellations in the zodiac. "*My path proceeds via 2 × 6 to the zodiac, since what is still older is always the newer,*" he concluded.

In the zodiac, the twelve signs were divided into four triangles of three signs. So perhaps the zodiac contained the solution as to how to get from three to four. Perhaps its symbolism, focusing on the quaternity, was more in touch with irrationality than was Christianity, which emphasized the Trinity. Followers of Christianity, he wrote, were "cosmic babies, 'greenhorns,'" their religion akin to the physics of people who "believe that the full moon and the new moon" are not the same moon.

Then he received a letter from his friend and one-time assistant, the physicist and Jungian disciple Markus Fierz. Fierz told him about a dream described by the sixteenth-century Cambridge Platonist Thomas More. In his dream, More glimpsed twelve sentences written in gold but he could recall only the first six. He was aroused from his dream by the braying of two donkeys.

Pauli set to work to analyze the dream. He reasoned that the splitting of the twelve sentences in More's dream into two sixes, along with the two donkeys, rendered "the quaternity" impossible. . . . The natural *splitting of 12 would be realized in 3 × 4*, which also corresponds to the splitting in the zodiac (of which Fludd always spoke). But the time was not ripe for a quaternary world-picture, and the dark half relapsed into the unconscious." In More's time, the zodiac must have been seen as two sixes—6 × 2. The world was not ready for the number four to emerge in Western thought.

It was a case of synchronicity. Pauli had homed in on 2 × 6, not realizing at first that he was reimagining the dreams of a man of the Renaissance, when these topics were hotly debated.

The division into two—symbolized by the two donkeys—reflected an era in which there was a split between matter and mind, soon to be manifested in the emergence of modern science in the work of Kepler and Descartes. Early scientists focused on matter because it could be described by mathematics in three-dimensional space, another case, like the Trinity, of three.

Or perhaps the two donkeys in More's dream had something to do with Pauli's own "two life phases." "A Zen Buddhist would understand this, as God speaks to us *always* in riddles," he added.

To illustrate the letter to von Franz, Pauli drew two six-sided stars of David, which he labeled as the two stages of his life: "the youth phase (until 1928) [and] the phase after World War II until the summer of 1953." The two stars sprang from Pauli's "active imagination"—two six-sided stars: 2 × 6.

In the years between these two life phases—1928 and 1945—Pauli lived through many turbulent events: two marriages, divorce, flight from Europe, rejection by colleagues at the ETH, the lonely war years in the United States, and his analysis by Jung, which had led to a deeper understanding of himself and the world about him. He depicts these two contrasting periods in three pairs of opposing archetypes.

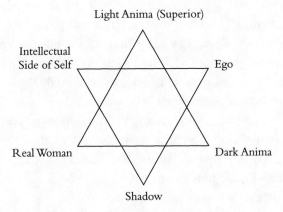

Light Anima (Superior)

Intellectual
Side of Self

Ego

Real Woman

Dark Anima

Shadow

The "youth phase" of Pauli's life, to 1928. (Based on Pauli's drawing in his letter to von Franz.)

In the youth phase of his life the archetypes are the "*Light Anima*" (superior; Pauli's intellectual side which was dominant in that "at the time I was completely absorbed in physics") versus the "*Shadow*, projected entirely on the father along with repressed negative emotions about Judaism"; the "intellectual side of the 'Self' projected onto the teacher" versus the "*Dark Anima* [whom Pauli saw as] ("inferior," a "prostitute"); and the "*Ego*" versus a "*Real woman* (mother) with whom I had a positive relationship."

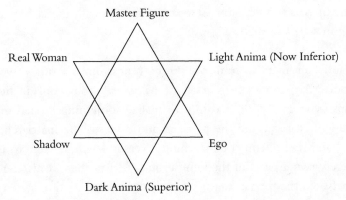

Master Figure

Real Woman

Light Anima (Now Inferior)

Shadow

Ego

Dark Anima (Superior)

The period of Pauli's life from 1946 to 1953. (Based on Pauli's drawing in his letter to von Franz.)

The transforming event in Pauli's life, as he remembered, was his marriage to Franca. This brought about the adult phase. Here the archetypes appear in new forms. The Light Anima, now inferior and projected onto the "wicked stepmother," is set against the Shadow. But the Shadow is no longer projected onto the father even though Pauli still regards him as "controversial, intellectual, divested of feeling." The "Ego" is set against the "*real woman*" with whom Pauli now has the "positive feeling of being at home"; this woman is no longer his mother. The third pair includes the "Master Figure" whom Pauli associated with Merlin. In an earlier letter to Emma Jung he described him as seeking a way to move from three to four solely by rational reasoning, using physics. But Pauli now knows this cannot be accomplished by logic alone and so places him in opposition to the Dark Anima who is now superior, appearing as the "Chinese woman."

Pauli then laid out the pattern of this, the second half of his life.

The connection with physics cools off. A positive connection with Kabbalah (and Chassidism) arises.

Then: the Kepler work.

Good and evil appear relativized and coincide with light (spiritual) and dark (chthonic).

Religion: is sought for.

Values of the youth phase threaten to be lost. At the high point of this phase a compensatory stagnation kicks in with anima-projections upon actual women.

Pauli continued to mull over three times four, six times two, and the zodiac. An "inner storm" raged. Finally his " 'active imagination' or perhaps 'synchronicity' " led him to realize something he had missed: "the dilemma regarding the patriarchalistic versus the matriarchalistic sphere [and the] dilemma concerning 3 versus 4—these seem to me to be merely two aspects of the same problem." In other words, the contrast between the patriarchalistic view (that is, the male-dominated view of the world in which the number three or Trinity predominates) and the matriarchalistic view (that is, female-dominated view of the world in which the quaternity predominates). According to ancient creation myths, the matriarchalistic world was overthrown and the patriarchalistic one took over.

Because "every 'correct' solution (i.e., meeting the demands of nature) must contain 4 as well as 3," Pauli tried different approaches including 16 (4 × 4) which can be mathematically tied to 12 (the number of signs in the zodiac). He explained why 12 and 16 were related: "The professor further calculates numbers. In the sign of the 12. 12:16 = 3:4—a problem with thorns and horns." Dividing each term in the ratio 12:16 by four gives the ratio 3:4.

But there was still something missing in the number twelve and its relationship with three and four, something Pauli could not quite put his finger on. He became weary and set aside the problem.

That night he had a dream. In his dream three others are present. One is a Chinese woman—now elevated to a "Sophia," a seer or wise woman. She tells Pauli, "You must play every conceivable combination of chess." Chess involves the opposition of black and white and thus of dark and light, the two animas. Dark is compassion and feeling, light is rationalism. It is like the teacher's advice in *The Piano Lesson*, that every possible melody can be played on the black and white keys: "It is only a question of knowing how to play." In his dream Pauli works on the problem of how to make four emerge from two, three, and six, but still can't solve it. He suddenly wakes up.

Then, in a waking vision, he hit on the solution. "This I will never forget. The experience had a numinostic character (obviously of archetypal nature)," he wrote in wonder, caught up in the moment of enlightenment. "In my youth I often had such experiences with physics problems," he added.

In the vision, a voice—possibly the Chinese woman's—wakes him from his dream. The voice tells him the solution in clear tones:

"In your drawings [the six-sided stars in Pauli's previous letter to von Franz] one element is perfectly correct and another transitory and false. It is correct that the lines number six, but it is false to draw six points. See here"—and I saw

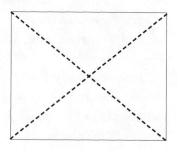

The square where the dance of diagonals occurred in Pauli's "waking vision."

a square with clearly marked diagonals. "Can you now see finally the 4 and the 6? Namely, 4 points and 6 lines—or 6 *pairs** out of 4 *points*. They are the same six lines that exist in the *I Ching*. There the 6, containing 3 as a latent factor, is correct. Now observe the square more closely: 4 of the lines are of equal length, the other two are longer—they are irrationally related, as you know from mathematics. There is *no* figure with 4 points and 6 *equal* lines. *For this reason symmetry cannot be statically produced and a dance results.* The coniunctio refers to the exchange of places during this dance. One can also speak of a game of rhythms and rotations.[†] *Therefore the 3, already contained in a latent form in the square, must be dynamically expressed.*"

Finally here was the solution to the problem that had been nagging at Pauli for so long. The four, or quaternity, emerged intact, while the three was also already contained "as a latent factor" in the square. This is the way it is in the *I Ching*, in which each hexagram used for divination is made up of six lines subdivided into two three-line segments (trigrams).

In mathematics the diagonals of a square cannot be expressed as whole numbers. Mathematicians refer to the relationship of the diagonals to the sides of a square as "irrational." Pauli extended the meaning of this term to psychology. In the dance of the diagonals, three and four are both present in a *coniunctio* resulting from "a game of rhythms and rotations" in which diagonals and sides transform into one another. Pauli recognized the influence of the *I Ching*—with its "6 lines and play of transformations"—on this solution.

---

[*]Pauli's footnote: "Obviously the combinations are from the previous dream.—Only a pair can play chess. (You see: the professor 'reckons.')"
[†]Pauli's footnote: "C.f. The chess game in the previous dream."

Perhaps what catalyzed this line of thought for Pauli was that the tetraktys plus its mirror image make up the holy quaternity. Two equilateral triangles fused along one of their sides form not a square but a parallelogram, while laying two equilateral triangles one on top of the other produces the star of David. Pauli's friend and former assistant, Markus Fierz—who was, we should remember, an acolyte of Jung's—had argued along these lines in a letter he wrote to Pauli a month earlier where he represented the opposition of light and dark each with its own tetraktys.

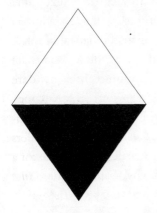

Fierz's geometric figure formed from two triangles, each a tetraktys.

The four-sided figure formed from the two triangles—one white, the other black—represents a state of the unconscious in which, Fierz wrote, light and dark "must either fly apart or flow one into the other," as the unconscious (represented by the dark lower triangle) flows into the conscious.

The other case in which three and four occurred simultaneously had been in Pauli's dream image of the world clock in which "three rhythms are contained." However, since the "image of the zodiac is not yet correct, then also is the 12 'incomplete,' " Pauli wrote to von Franz. The zodiac is "not yet correct"—he felt, but didn't know why. Perhaps the error lay in something Jung had missed in examining why the patriarchal view arose in Christianity: he had limited his clues to those within Christianity. It occurred to Pauli that the Zodiac had pre-Christian roots. He decided to look into the cultural history of something that Jung and his circle "too strongly neglected"—horoscopes. Even though Pauli had nothing but disdain for them, he saw their importance as a cultural artifact.

He discovered that the horoscope then in use was based on the zodiac of third-century-B.C. Babylonia. But this was a time when patriarchy had replaced matriarchy. So this zodiac is associated with the all-male Trinity of Father, Son, and Holy Ghost which, being three, is incomplete. The use of this patriarchical horoscope had much to do with the masculine slant of Christianity and so, too, its preference for the number three over four. Jung had been wrong to focus his attention on the Christian era in his search for the emergence of four from three.

Thus Pauli resolved the 2,000-year-old problem of how to go from three to four—to his own satisfaction, at least. In fact, he realized that the problem was more than 2,000 years old, harking back to Babylonian times and the venerable *I Ching*. Pauli's liking for mathematical symmetry carried over to the *I Ching*, which, as he put it, exhibited "the exact *symmetrical* mental attitude of the pairs Yin (feminine, chthonic, dark = moon) and Yang (masculine, intellectual, light = sun)." Horoscopes, however, exhibited no such symmetries.

In Pauli's opinion horoscopes "*far exceed all rational thought*" and so had no value at all. But the *I Ching* with its properties of synchronicity appealed to him "instinctively." And he now knew, thanks to Jung, that intuition was the psychological function best suited to take in the whole situation.

Pauli did however once have a horoscope drawn up and included it in a letter he wrote to Jung in December 1953. It was not published in the Jung/Pauli correspondence and the two never discussed it in their letters. It is not known who constructed the horoscope and it is unlikely that Pauli attached much significance to it. Yet it may or may not have had something to do with Pauli's statement to Jung that equinoxes were times of "relative psychic instability, which can manifest itself both negatively and positively (creatively)." In Pauli's horoscope the spring equinox (the boundary between Pisces and Aries) is on the cusp between the seventh house, the "house of conjunction," and the eighth, the "house of the unconscious" and the autumn equinox is on the cusp between the first house, the "house of the ego" and the second, the "house of material things." One interpretation of Pauli's instabilities might be that they reflected the instabilities of these boundaries between houses.

In working on the psychological side of the problem of how we go from three to four Pauli had gained a deeper understanding of himself and how the whole course of physics seemed to be guided by the "arche-type of the quaternity." He saw this course as leading to an extension of archetypes out of the collective unconscious into a new form of phys-ics. This would surely be a major result of his joint work with Jung. He wrote to him:

> In this way, the ancient alchemical idea that matter indicates a psychic state could, on a superior level, experience a new form of realization. I have the impression that this is what my physical dream symbolism is aiming at.

PAULI had definite ideas on how an appropriate form of mysticism would appear. He was adamant that "my real problem was and still is the *relation between Mysticism and Science*, what is different between them and what is in common. Both mystics and science have the same aim, to become aware of the unity of knowledge. . . . And who believes that our present form of science is the last word in this scale? Certainly not I."

In the summer of 1957 Pauli described his attitude to mysticism in a letter to the Israeli physicist and historian of science, Shmuel Sambur-sky: "In opposition to the monotheist religions—but in unison with the mysticism of all peoples, including Jewish mysticism—I believe that the ultimate reality is *not* personal. Thus is it also in the Vedantic philosophy, and so it is in Chinese Taoism, the Nirvana of Buddhists . . . and the En-Sof of Jewish mysticism. It is the task of mankind, through personal asso-ciation not to implement these forms themselves (Yoga-teaching). . . . In this sense only is *Yahweh* for me a local demon who displays his efficacy primarily in Israel. How has he behaved with me? He was relatively mild, he only beat me gently on the left ear." Why "left"? Perhaps Pauli was thinking of his unconscious. Or perhaps he meant that God gently chided him for suggesting the existence of the neutrino, the weak par-ticle that ended up causing a revolution in physics because it spins only to the left.

As for von Franz, in 1955 Pauli was writing to her using the familiar "Du." But two years later their exchanges abruptly ended. Perhaps Pauli felt he opened up too much to her. As he once wrote to her, when it comes to feelings "there I am no celebrity, but underdeveloped, even infantile." Von Franz claimed that he had become unpleasant and wanted her to analyze his dreams for free. She was bitter about this and, in turn, became critical of him.

According to Charles Meier, who knew both of them, "she totally misunderstood Pauli, failing to appreciate his efforts to conduct an analytic dialogue with her and that their relationship was tragic."

A supreme example of synchronicity: The Pauli effects pile up

On May 26, 1955, Pauli was due to give a lecture on Einstein at the Zürich Physical Society, to celebrate the fiftieth anniversary of the discovery of the special theory of relativity. Before the lecture, three of his friends and colleagues met up for a teetotal dinner. Then they all set off for the meeting.

David Speiser, a young Swiss physicist, was on his Lambretta. He discovered he was low on gas and went to a gas station. Then his scooter suddenly caught fire. He threw a jug of water over the flames and extinguished them but the scooter was totaled and he had to walk. Another young Swiss physicist, Arman Thellung, discovered his bicycle had two flat tires so he also had to walk. Ralph Kronig, the original discoverer of electron spin, took the tram. It was a journey he had made many times in the past but for some reason he failed to notice the Gloriastrasse stop and forgot to get out. He only realized several stops later. It was a magnificent example of a multiple Pauli effect. Fortunately they all arrived on time for Pauli's talk and Pauli himself was most amused to hear about their mishaps. As Thellung recalled, "a defining feature of the Pauli effect was that Pauli himself never experienced any harm."

Fierz wrote: "Pauli himself thoroughly believed in his effect. He has told me that he senses the mischief already before as a disagreeable tension, and when the anticipated misfortune then actually hits—another one!—he feels strangely liberated and lightened. It is quite legitimate to understand the 'Pauli effect' as a synchronistic phenomenon as conceived by Jung."

13

Second Intermezzo—Road to Yesterday

The dream

ON NOVEMBER 4, 1955, Pauli's father died. By now the two had reconciled and Pauli was deeply affected. It felt like a defining episode in his life. As he put it, "the shadow with me was projected onto my father for a long time, and I had to learn gradually to distinguish between the dream figure of the shadow and my real father."

Despite the rift between them after the suicide of Pauli's mother, Wolfgang Sr. had been always immensely proud of his son's achievements. Eventually father and son managed to overcome their differences and were back on good terms. Perhaps their reconciliation was a combination of time healing emotional wounds and, of course, Jung's therapy.

When Germany invaded Austria, in 1938, it placed Pauli's father in great danger. For despite his conversion, he had been born a Jew. Pauli immediately arranged for him to move into Switzerland. Wolfgang Sr. had to leave all his possessions behind and arrived in Zürich with only a suitcase, accompanied by his wife, Maria (Pauli's "wicked stepmother"). Initially they stayed with Pauli and Franca. Franca did not get along with Maria. In the end Maria decided to return to Vienna and was not reunited with her husband until after the war. In Switzerland Wolfgang Sr. was welcomed at the chemistry department of the University of Zürich, where he continued his scientific research. After he died Maria had severe monetary problems, often turning to her stepson for help. She also had an alcohol problem.

Soon after Pauli's father's death, Jung's wife Emma also died, followed by the elegant mathematician Hermann Weyl. Weyl's cremation was set for December 12 at 17.00 hours.

A few weeks earlier, on October 24, Pauli had had a dream. In it he is on an express train, departing at 17.00 hours—the exact time of Weyl's cremation. The train encounters an obstacle and has to swerve around it. Then Pauli goes into a church with Franca and a Swiss friend whom he calls Mr. X. In the church "some strangers" are waiting—the strangers who occur again and again in Pauli's dreams. There is a blackboard in the church. Pauli goes up to it and writes some complicated equations to do with the quantum theory of magnetism.

Then a famous preacher appears, the "Master" or the "great stranger." He walks to the blackboard and says in French that the subject of that day's sermon *"will be the formulas of Professor Pauli. There is here an expression with four quantities,"* taken from one of Pauli's equations on magnetic effects. The equation reads: $\mu HN/V$. Mu (μ) is the extent to which a material is magnetized, H the magnetic field produced by the number (N) of electrons in the magnet, and V is the magnet's volume.

In all, there are four symbols—the number four again. Later Pauli recalled that in Jung's books, particularly *Aion*, he had mentioned that magnets were often considered a source of magical power, in that they contain opposite poles, north and south, in a single object.

In Pauli's dream the strangers become excited and shout in French *"parle, parle, parle"*—"speak, speak, speak." As always they want him to speak about feelings (France being the country of feelings) and about physics and psychology. (In his account of this to Jung, Pauli comments humorously, "In my dreams, by the way, I often speak somewhat better French than I do when I am awake.") But he is reluctant to speak up, fearing for his reputation among fellow scientists. His heart begins to pound so hard that he wakes up.

Musing over his dream, Pauli interprets the church as a new house, free from any struggle between opposites. Franca is with him. He is at one with himself.

Through analysis, he tells Jung, his function schema has changed. In earlier years his thinking function was dominant, but now that role has shifted to his intuition. Things are going better with his feeling side— represented by France—while extraverted sensation has become the inferior function. In other words, Pauli has become a nicer person, though further removed from reality.

This self-assessment was corroborated by Marie-Louise von Franz, who said of him: "He was highly intelligent, very honest in his thinking, but otherwise a very immature big boy in his feelings. . . . He had a patriarchal outlook on women. Women were pleasant things to play with, but not something to take seriously."

Pauli described his dream to Fierz as well as to Jung. Fierz asked rhetorically, "To where is this journey?" He pointed out that the formula Pauli had written on the blackboard referred to optics as well as magnetism. The combined subject is called magneto-optics and concerns how light is transformed when it is passed through a material immersed in a magnetic field. Pauli had made important advances in it, one of which was this formula. Fierz reminded Pauli that magnetism was to do with attraction—the attraction of the north and south poles—while optics refers to visualizations. "What is the connection for us of these magnetic visualizations?" he asked.

"The connection," he continued, answering his own question, "is an alchemical one which concerns a transformation leading to an unfolding of events. How so and why so, you know much better than others from personal experience. 'The magneto-optical transformation and the 4 quantum numbers,' this is the key to your biographical experience." But Fierz did not know the meaning of the journey in the dream or how it related to the alchemical notion of transformation as Jung reinterpreted it.

In fact Pauli was about to make a very significant train journey.

Brief encounter

Pauli told the story of his journey in three different ways in three different letters: to his friend Paul Rosebaud, the Scientific Director of

Pergamon Press; to Fierz; and to Jung. The events took place during a trip to Hamburg between November 29 (full moon night) and December 1. Pauli also described the trip to Bohr. He referred to it only as "this 'road to yesterday,'" with no details.

Pauli's trip to Hamburg on November 29 was to give a lecture at the university there on "Science and Western Thought." He spoke on how important it was to reconcile the rational-critical (that is, Western science) with the mystical-irrational (that is, Eastern thought) to try to create a single framework of the physical and the psychical. It was an important lecture for him because it was one of the very few times he ever spoke in public on this topic. "It is precisely by these means," he concluded, "that the scientist can more or less consciously tread a path of inner salvation. Slowly then develop inner images, fantasies or ideas, compensatory to the external situation, which indicate the possibility of a mutual approach of poles in the pairs of opposites."

Early that evening, at precisely 17.00 hours, the phone in his hotel room rang. Pauli picked up the receiver. He recognized the voice immediately. It was the beautiful, blonde girlfriend from his Sankt Pauli days whom he had abruptly dumped when she had become a morphine addict. In writing about this meeting to his friends, he kept the woman's name a secret.

Ten years earlier, she said, she had seen his name in a newspaper, announcing that he had won the Nobel Prize. But she couldn't track him down. She didn't know where he was living. Then she spotted an advertisement in a newspaper saying that he was to give a lecture in Hamburg. His hotel was also named.

Pauli was excited to hear from her and curious as to what had become of her. But he was also apprehensive. Even though so many years had passed since then, he still felt the old dread of mixing his night and day selves. In the end he agreed to meet her on the day he was due to leave, in the lobby of his hotel, the Hotel Reichshof, one of the best in Hamburg.

She was two years younger than he, so she would have been fifty-three. Nevertheless, when she came through the revolving doors he saw that she was still beautiful, blonde, and alluring. "This young woman suddenly appears *qua* woman. (Regeneration motif! N.B. shortly before,

on 4 November, my father died) and she was in good health," he wrote
to Fierz.

For two hours they talked intensely. "A whole lifetime of 30 years
passed before me—her cure, a marriage, and a divorce, with war and
National Socialism as a historical background."

Perhaps, he thought, the situation was archetypal, a fairytale "being
played out." After all, November 29, the night that she contacted him,
was the night of the full moon. The first time they met, he recalled, had
been in his Jekyll and Hyde days in Hamburg:

> 30 years ago my neurosis was clearly indicated in the complete split
> between my day life and my night life in my relations with women,
> but now it was very human.

As for seeing her again—"erotic it was not." Rather it was a painful
reminder of the very different person he had been thirty years earlier:

> I saw myself as in a dark mirror, in a time 30 years previous with
> its sharp cut between the worlds of night and day, which I worked
> strenuously to maintain—until the breakdown of 1930 came upon
> me (my great life crisis). In the day calming works, in the night
> sexual entertainment in the underworld—without feeling, without
> love, indeed without humanity. "What price glory!"

Pauli told her that he regretted falling out of contact. He recalled
how pretty she had been in those days and, he added, now too. A human
life unrolled before him. He felt ashamed that he had no more to say.
"But should I speak—'speak, speak'—what should I say," he thought, in
French. As with the "strangers" in his dream who wanted him to speak
about his feelings, he was tongue-tied in front of this woman.

"One should not give up hope," he wrote to Fierz. "I gave up on
this young lady in 1925, *too early*. But, personally, even if she at the time
would have been entirely healthy, the life rhythms were not so. (It is very
probable, that at that time I kept away from her—I wanted at that time
no external connection [no relationship])."

The woman walked Pauli to his train. When they arrived they were both aware that this was a moment that would not occur again. Pauli told her how pleased he was to have seen her again. He was married, he said, lived in Zürich, and was easily within reach. He thought of kissing her, then hesitated and decided not to. "Now it was as friends—at that time *not*."

He described their parting to Jung:

> But now it was very human, and as we parted on the platform, it seemed to me like a *coniunctio*. Alone in the express train to Zürich, my mind went back to 1928 as I took the same route toward my new professorship and my great neurosis. I may be a little less efficient than in those days, but I think the prospects are a bit brighter as regards my mental and spiritual well-being.

Afterward Pauli could not put the meeting out of his mind. In search of a deeper understanding of it he thought back to Fierz's comment on his dream about the blackboard. Fierz had pointed out the fourness it contained in the four symbols in Pauli's equation as well as the four quantum numbers. All of a sudden, Pauli realized that this research was part of his tumultuous past, his Hamburg days. The blackboard dream had occurred before the death of his father, a defining episode in his life. Meeting the woman in Hamburg made him realize that he had broken with that past. "There was a transformation in those 30 years, at the [station] platform [where they said goodbye] it had for me somewhat of a humanistic-conciliatoriness, a '*coniunctio*.' . . . My individual life attained a sense of symmetry between past and present," he wrote to Fierz. Just as Fierz had suggested, the number four held the key to his "biographical experience." Pauli's train ride was not to meet his neurosis but to return home.

"May sometime the preacher of dreams make it possible for you to make a journey as I have made to Hamburg," Pauli finished his story.

Through the Looking Glass

Dreams of reflections

IN NOVEMBER 1954 Pauli had a dream so curious that three years later he was still thinking about it. He sent a description of it to Jung. In his waking life Pauli was very preoccupied with issues of symmetry, both in physics and in psychology—the conscious and unconscious as mirror images of each other. It was not surprising that this preoccupation should seep into his dreams as well.

In the dream he is with a dark woman—his anima—in a room in which experiments are being carried out involving reflections. The "others" in the room think that the reflections are real objects, but Pauli and the dark woman know they are just mirror images. They keep this secret. "This secret fills us with *apprehension*." From time to time the dark woman changes into the Chinese woman of Pauli's earlier dreams, who had paraded Pauli before the "strangers." The Chinese woman, according to Jung, represents the dark woman's holistic side, in that Chinese philosophy seeks to reconcile opposites.

Pauli guessed that the "others" represented the collective opinion which he took to be his "own conventional objections ... to certain ideas—and [his] fear of them." The problem with which he is struggling is that "there is *no* symmetry of 'objects' and 'reflections' in this dream, since the whole point is about distinguishing between the two." Even though he can see that what appear to be objects are simply reflections, the "others" cannot. In this dream there is no mirror symmetry. But this is impossible.

In the very first set of dreams he had presented to Jung in 1932, he had spoken of the conscious and the unconscious as mirror images of each other. When he was sure that the "left is the mirror image of the right," he felt at one with himself because it meant his conscious and unconscious were in balance.

*Charge, parity, and time reversal (*CPT*)—Pauli's third great breakthrough*

Pauli never worked out quite why it was that he started working on the mathematics of mirror images. The year was 1952. As he wrote, "between 1952 and 1956 there was not actually anything going on in the world of physics to justify focusing on that particular subject." It was two years later that he dreamed about the Chinese woman and began to suspect that "there must have been psychological factors involved."

Pauli began his investigation into mirror symmetry by looking at time reversal. Representing time with the symbol T, he placed a minus sign in front of it. In mathematical terms, he made time run backward. (Hard though it is to imagine, this can actually be done in the laboratory at a subatomic level.) When time is run backward, we look into a world in which a particle that was originally moving to the right is now moving to the left; the particle's speed is also reversed. The law of time reversal asserts that the laws of physics remain the same when time is reversed—time-reversal invariance—which means that there is no difference between the state of a collection of moving elementary particles (or billiard balls or cannonballs) and the same collection in which time has been reversed. Time-reversal invariance is not a property of a specific state of a collection of objects, but of two different states. In other words, it is not intrinsic to any particular particle, like the spin of an electron is.

Pauli's lectures on this inspired other physicists to look into the connection of time reversal (T) with two other symmetries, parity (P) and charge conjugation (C). Parity is the mirror-image effect, the interchange of left and right. In both prequantum and quantum physics, the law of parity conservation states that every physical system (that is, atoms, electrons, billiard balls, etc.) should be independent of any difference between right and left. Physicists believe that atomic systems have a particular parity independent of the system's location in space and time and which must be maintained throughout all interactions. If an experiment is performed it should turn out exactly the same when observed in a mirror; it should be a perfect mirror reflection. This is called parity invariance (P-invariance) or mirror symmetry.

In mathematics, the law of parity conservation means that a physics equation must remain the same when right and left are exchanged. Parity invariance explains many properties of atomic systems and, at the time that Pauli began working on it, was taken as axiomatic; it had never been questioned.

By the mid-1930s physicists agreed that every charged particle must have a matching antiparticle. Charge conjugation (C) describes the mathematical process used to convert every particle into its antiparticle, and so matter into antimatter. In quantum mechanics it is considered to be an intrinsic property of every elementary particle (for example, an electron) along with parity.

In 1954, two years after he began working on mirror images, Pauli decided to look deeper into the whole subject. Instead of exploring charge conjugation (C), parity (P), and time reversal (T) separately, he looked at all three together—the whole combined operation of CPT. This entails the exchange of particle and antiparticle (C), right-left symmetry (P), and time reversal (T)—(particle $\leftrightarrow$ antiparticle) x (left $\leftrightarrow$ right) $\times$ (future $\leftrightarrow$ past). CPT makes the astonishing assertion that a mirror universe—in which all matter is replaced with antimatter, all positions are their own reflections, and even time runs backward so that all speeds are reversed—would actually be indistinguishable from our own.

Explaining this complex concept to Jung, Pauli wrote, "The combination of CPT of all three parity operations . . . is correct under much more general assumptions (that is, deducible, demonstrable) than the operations $C, P,$ and T taken individually." The exclusion principle played a central role in his calculations because it required that collections of particles and antiparticles with half a unit of spin and those with a whole number of spin be treated differently. He proved that if the equations describing an atom remain unchanged under *CPT*, this was the same as requiring any of those equations to agree with relativity theory.

To date no violations of *CPT* symmetry have been found in the laboratory—that is, incorrect predictions of equations which can be traced to violating *CPT*, or what is the same, with violating relativity theory. It underlies every theory of elementary particles and is essential for studying the behavior of its three component symmetries—*C, P,* and *T*—in this increasingly bizarre world.

Pauli wrote up his conclusions in a milestone paper to be published in a volume to celebrate Niels Bohr's seventieth birthday the following year, 1955. It was the capstone to a subject he had studied with a passion ever since he was a young man—relativity—and it finally set his greatest discovery, the exclusion principle, firmly in its embrace, along with another offshoot of the exclusion principle, spin. He began his paper on a historical note, recalling "a long and still continuing pilgrimage since the year 1922, in which so many stations are involved."

It was precisely as he was writing this paper that he dreamed that curious dream about the Chinese woman and the reflections that were not reflections. But at the time he thought no more about it.

The father of the neutrino

In June 1956, the experimental physicists Frederick Reines and Clyde Cowan made an exciting discovery. They succeeded in finally verifying in the laboratory the neutrino, which Pauli had predicted twenty-six years earlier, actually existed. They immediately sent him a telegram. Pauli was at a symposium in CERN, the huge nuclear physics laboratory outside Geneva, at the time. Full of elation he read it out to the audience: "We are happy to inform you that we have definitely detected neutrinos from fission fragments from observing inverse beta-decay of protons. Observed cross-section agrees well with expected 6×10^{-44} cm^2."

Pauli was acclaimed worldwide as the "father of the neutrino" and was much called upon to give lectures on the subject. Legend has it that he commented, with quiet satisfaction, "Everything comes to him who knows how to wait."

The downfall of parity

That same month—June 1956—two Chinese American physicists, T. D. Lee and C. N. Yang, sent Pauli an article they had written in which they argued that perhaps mirror symmetry—parity—might not always be conserved. They had studied the scientific literature and were convinced there was very little actual experimental evidence to support it, added to which certain puzzling phenomena in elementary particle physics could be clarified if it was considered to be not generally valid. They suggested specific experiments for testing their proposal. To suggest that the law of parity might not be inviolate was outrageous. Pauli chuckled and put the article aside.

Nevertheless the two presented such a powerful case that other physicists became curious and started carrying out experiments. On January 17, 1957, Pauli wrote to Weisskopf that he was "ready to bet a very high sum" that the experiments would fail. Little did he know that the previous day *The New York Times* had carried a front-page report on what it called the "Chinese Revolution" in physics. A group of physicists from Columbia University headed by a woman, Chien-Shiung Wu, had carried out a very beautiful experiment proving beyond a doubt the overthrow of parity in the case of weak interactions.

Weak interactions are interactions between elementary particles—such as electrons, protons, and neutrinos—occurring with a force far weaker than the nuclear force, which binds the nucleus together, or the electromagnetic force between charged particles. On a scale of one to ten, the nuclear force is magnitude 1; the strength of the electromagnetic force, given by the fine structure constant, is 0.00729—which can also be expressed as 1/137;* while the weak force is a mere 0.00000000000001, very weak indeed.

---

*As well as describing the DNA of spectral lines, 1/137 also determines the strength of the interaction between two electrons, because the charge of the electron occurs twice (e^2) in the fine structure constant equation, $2\pi e^2/hc$.

In their experiments Wu and her co-workers monitored the number of electrons emerging from the nuclei of a radioactive isotope (an alternative form) of cobalt undergoing beta-decay. The beta-decay process involved the transformation of a neutron in the nucleus of the radioactive sample of cobalt into a proton, electron, and neutrino. To define a direction—in this case up and down—they aligned the spins of the nuclei by placing them in a magnetic field. If parity really was a universal law, if nature really did not distinguish one direction from another, then one would expect precisely equal numbers of electrons to be emitted upward and downward. But this was not what happened. In fact, the electrons came out asymmetrically. In other words, Wu's experiment was not the same as its mirror image. The beauty of this experiment—which scientists found so impressive—was its precision. The apparatus had to be set at the lowest possible temperature, close to absolute zero. This was necessary to eliminate any movement of the nuclei due to heat agitation, which would have ruined the alignment of their spins. Thus Lee and Yang were proved to be right.

So parity had been overthrown in the weak interactions—an event which struck Pauli like a bolt of lightning. He was, he confessed "very upset and behaved irrationally for quite a while." He wrote to Weisskopf that he was glad that in the end he had not made any bet. "It would have resulted in a heavy loss of money (which I cannot afford); I did make a fool of myself, however (which I think I can afford to do)," he said.

He wrote humorously to Bohr about the event (a slip of the pen led him to use the wrong date for the overthrow of parity):

It is our sad duty to make known that our dear female friend of many years

PARITY

had gently passed away on January 19, 1957, following a brief suffering caused by experimental treatment. On behalf of the bereaved.

$$e, \mu, \nu.$$

(e, μ, ν—electron, muon, and neutrino—are three of the many particles that participate in the weak interactions.)

In real life, of course, our bodies are not symmetrical; our heart is on the left, for a start. But up until then scientists had taken for granted that the laws of physics were mirror symmetrical. Equations had always been drawn up on that assumption. At the atomic level, at least, it had turned out that this was not invariably the case. Perhaps "nature is not mathematical and does not conform to our thinking," Pauli wrote to Fierz. It was as dramatic a revelation as the Pythagoreans realizing that the square root of two was not a whole number.

Pauli firmly believed that principles of symmetry had to prevail. The way to find them lay not only through logic, but in the more irrational dimensions of thought. "With me the mixture of mysticism and mathematics, which finds its main results in physics, is still very dominant," he wrote.

Two decades earlier Pauli had argued with Bohr when Bohr suggested that the law of the conservation of energy might not entirely hold in the case of beta-decay. It was to preserve this law that Pauli had proposed the existence of the neutrino. As we have seen, the neutrino had been discovered in the laboratory six months earlier and played a role in Wu's experiment. It is one of the most weakly interacting particles of all.

"[Bohr] was wrong with the energy law, but he was right that the weak interactions are a very particular field where strange things could happen, which don't happen otherwise," Pauli wrote. He added that physicists should bear in mind something else Bohr had said: "We have to be prepared for surprises." Pauli himself was willing to speculate that in interactions much weaker than the weak interactions, not just parity but energy too might not be conserved (that is, the amount of energy involved at the start of a process is not the same as at the end). " 'Be prepared for surprises' not anywhere but specifically with the beta-decay," he wrote.

Pauli had met Chien-Shiung Wu in 1941 when he visited the University of California at Berkeley and described her to Jung as impressive, "both as an experimental physicist and an intelligent and beautiful Chinese young lady." Born in Shanghai in 1912, she had come to America in 1936 and worked first at Berkeley before moving to Columbia. Photographs show her formidable intelligence as well as her beauty. She was a perfectionist—just the person needed to attempt the highly precise experiment to test parity.

Pauli wrote to her that "what had prevented me *until* now from accepting this formal possibility [of parity violation] is the question why this restriction of mirroring appears only in the 'weak' interactions, not the 'strong' ones." But what did the strength of an interaction have to do with a law of conservation? The question has still to be answered. "In any case, I congratulate you (to the contrary of myself)," he added. "This particle neutrino—on which I am not innocent—still persecutes me."

"God *is* a weak left-hander after all," he wrote exuberantly to Jung. A neutrino only spins in one direction. If one looked at it in a mirror, it would still spin in the same direction. Most screws are right-handed—you twist the screwdriver in the same direction as the curl of the fingers on your right hand and turn the screw toward your right thumb. Neutrinos are left-handed in that they spin in a direction opposite to their motion, the direction in which you would twist a screwdriver for a left-handed screw. So God is a left-hander, but a weak one, because he is only a left-hander in the weak interactions—featuring the neutrino—where parity is violated.

The Chinese woman

Pauli could not fail to notice what a supreme example of synchronicity this was. A Chinese woman had played an important part in his dreams, particularly those involving mirrors and their reflections; and a Chinese woman had carried out the critical experiment that brought about the downfall of parity—that is, of mirror symmetry—in physics. He wrote to Jung of his "shock" at this " 'Chinese revolution' in physics."

Fierz told him he had "a mirror complex." "I admitted as much," Pauli wrote to Jung. "But I was still left with the task of acknowledging the nature of my 'mirror complex.' "

Pauli's curious dream of 1954 had occurred right after he finished his work on mirror symmetry. He was convinced that "unconscious motives play a role" in creative thinking, especially in the case of symmetry. " 'Mirroring' is an archetype [and] this has something to do with physics. Physics relies on a connection of an image reflected in a mirror and between mind and nature," he said in an interview in 1957. He recalled having "vivid, almost parapsychological dreams about mirroring, while I worked mathematically during the day." The mathematical work seemed to cause "some archetype [to be] constellated [that is, to emerge into consciousness] which subsequently made me think about mirroring." The connection, he concluded was "a kind of synchronicity, because there are unconscious motives when one is involved in something."

Other examples of synchronicity soon cropped up. Two months after the ground-breaking experiment that spelled the end of parity, in March 1957, Pauli's friend, Max Delbrück, an eminent biologist, sent him an article on a one-cell, light-sensitive mushroom known as a *phycomyces*.

A few weeks later Pauli was talking about psychophysics with Karl Kerényi, an authority on Greek mythology and a close friend of Jung's. Pauli told him about some dreams he had had in which he was wandering about in the constellation of Perseus. Synchronisms started to spring up. For a start, Perseus contains a binary, or double star, known as Algol. Moreover, in Greek mythology, Perseus fought Medusa while looking at her reflection in a mirror. Shortly afterward Pauli came across an article by Kerényi on Perseus. It concluded with an ancient Greek pun about Perseus's founding of the city of *Mycenae*. It had been so named after a mushroom called *myces*—the very mushroom that had been the subject of Delbrück's scientific paper.

Mirror images

That same month—March 1957—Pauli had a dream in which a "youngish, dark-haired man, enveloped in a faint light" hands him a manuscript. Pauli shouts, "How dare you presume to ask me to read it? What do you think you are doing?" He wakes up feeling upset and irritated.

Writing about the dream to Jung, Pauli suggested it revealed his "conventional objections to certain ideas—and my fear of them," most notably his belief that parity could never be violated.

In another dream a couple of months later he is driving a car (though in real life he no longer had one). He parks it legally but the young man from the previous dream suddenly appears and jumps in on the passenger side. He is now a policeman. He drives Pauli to a police station and pushes him inside.

Pauli is afraid that he will be dragged from one office to the next. "Oh no," says the young man. An unfamiliar dark woman is sitting at the counter. In a brusque military voice the young man barks at her, "Director *Spiegler* [Reflector], please!" Taken aback by the word "Spiegler," Pauli wakes up. When he falls asleep again the dream continues.

Another man comes in who resembles Jung. Pauli assumes he is a psychologist and explains to him in great detail the significance of the downfall of parity on the world of physics.

For Pauli the dream reflects his long-held belief that the "relationship between physics and psychology is that of a mirror image." In the dream he appears first as his narrower Self who understands both physics and parity, then as his own mirror image—the psychologist who knows nothing about either. Spiegler—the reflector—is responsible for bringing out the psychologist and is attempting to bring the two together. But now that parity has been violated, there is no longer any mirror symmetry.

Looking at the question in terms of archetypes Pauli finds the loss of mirror symmetry not so shocking. Before the downfall of parity, he feels physicists and psychologists had not been looking deeply enough into matter and mind. They had considered only "partial mirror images." Full reflections and more profound symmetries can be obtained only by going deeper into the psyche. The *CPT* symmetry that Pauli himself discovered—exchange of particle and antiparticle, symmetry of right-left, and time reversal—is exactly that profound symmetry because it talks about mirror symmetry on the grandest of scales. It asserts that our universe cannot be distinguished from a mirror universe in which all matter is replaced with antimatter, all positions are reflections, and time runs backward.

Thus Pauli worked out that parity could be restored in a new and profound way by taking into account a fuller symmetry—*CPT*, which reveals the full symmetry of phenomena.

Over the years scientists have discovered some of the stunning implications of Pauli's *CPT* symmetry. One is the following. In experiments on certain elementary particles it seemed that the combined symmetry of *CP* (matter/antimatter and parity) was violated. As one was violated and the other not, the two together—the product $C \times P$—is violated. (As in mathematics where $+1 \times -1 = -1$.) This is therefore a loss of symmetry.

For *CPT* to remain valid, time reversal (*T*) would have to be invalid too, which would make the combined symmetry of the three—*CPT*—valid. (As in −1 × −1 = +1.)

In the late 1990s, scientists actually produced direct proof of the violation of time-reversal invariance on the subatomic level, that is, when time was made to run backward the laws of physics concerning this specie of weak interaction did *not* remain the same. From this they were able to show that the transformation of matter into antimatter is not symmetrical in time. The cosmic implications of this are enormous. It helps explain a question that intrigues physicists: why the universe is made up of matter rather than antimatter, even though equal amounts of both were created in the big bang.

After the fall of parity, as Pauli commented with quiet pride to Jung, "The '*CPT* theorem' was on everybody's lips." "To many physicists *CPT* was a fixed point around which all else turned," T. D. Lee recalled of that turbulent era. It also seemed to be a way to bring together Pauli's interests in physics and psychology. He had "no doubt that the placing side by side of the points of view of a physicist and a psychologist will also prove a form of reflection." The startling "mirror" symmetry of *CPT* related elementary particles in a new and profound way. So why should the apparently dissimilar views of a physicist and a psychologist not mirror each other as well?

Pauli and T. D. Lee (whose theoretical work had brought about the downfall of parity) quickly developed a rapport. Pauli was intrigued by Lee's research on how elementary particles transform into one another. On one occasion Pauli visited Lee at the Brookhaven National Laboratory on Long Island. They planned to go out for dinner that evening with their wives. There was valuable scientific equipment on site and a guard was posted at the gate. As he was leaving in his car, Lee handed the guard his identification card. An uncomfortably long time went by. Lee inquired whether there was a problem. The guard apologized. He had somehow misplaced the card, he said. In all his years on duty it was the first time it had ever happened. Lee laughed. It was a first for him, too. The guard finally located the card; it had fallen through his fingers under a table. Pauli exclaimed gleefully from the back seat, "It's the Pauli effect!"

From mirror symmetry and archetypes to . . . UFOs

The violation of parity—that it was, in fact, possible to distinguish between left and right in atomic physics—struck a chord in Jung's ongoing fascination with unidentified flying objects (UFOs), an interest which, as Jung admitted, "might strike some people as crazy." At the time—the 1950s—many people were fascinated by UFOs and a number of highly successful books had been written. During their long dinners at Jung's huge house on the lake, Pauli and Jung often discussed the subject. Jung begged Pauli to make inquiries about flying saucers among his scientific colleagues. Pauli's theory was that they were either hallucinations or secret experimental aircraft invented by the Americans. He had nothing but scathing comments on the many manuscripts that were sent to Jung on the subject.

Taking up Pauli's quip on the violation of symmetry—that "God *is* a weak left-hander, after all," Jung declared that "the statements from the unconscious (represented by UFO legends, dreams, and images) point to . . . a statistical predominance of the left—i.e., to a *prevalence of the unconscious*, expressed through 'God's eyes,' 'creatures of a higher intelligence,' intentions of deliverance or redemption on the part of 'higher worlds' and the like."

Jung was convinced that at this moment in history the unconscious was in a stronger position in relation to the conscious—a dangerous situation. The way to resolve this imbalance was through the redeeming Third—an archetype of some sort or other, a latent symbol. The UFO legend was perhaps this latent symbol. It was lurking somewhere in the psyche where it was trying to elevate the collective unconscious to a higher level. This would ultimately resolve the conflict of the unconscious and the conscious by paving the way for a dialogue between them, finally permitting the Self to emerge in the process Jung called individuation. The Third provides asymmetry and tips the balance toward the conscious. Putting it in physics terms, Jung thought that certain key elementary particles in the weak interactions played the part of the Third, while the law of parity of an object and its reflection paralleled the opposition between conscious and unconscious and right-wing and left-wing in the political sense.

Jung saw "an almost comic parallel" in the tumult in physics caused by the weak interactions. It was precisely the same as when tiny psychological factors "shake the foundations of our world." "Your anima, the Chinese woman," he wrote to Pauli, "already had a scent of asymmetry."

For Jung the UFO legend indicated that the Self was ultimately *Spiegler*, "The Reflector," representing both a mathematical point and the circle, universality, "God and mankind, eternal and transient, being and nonbeing, disappearance and rising again, etc." "There is absolutely no doubt," he concluded, "that it is the individuation symbolism that is at the psychological base of the UFO phenomenon."

All the same, as far as Jung was concerned, UFOs were not merely symbols but very real, as evidenced by the many historical sightings recorded in the books, magazines, and newspapers piled high in his study.

After one of their lengthy late evening conversations on the subject of UFOs, Pauli had a sighting, though not of a UFO:

> As I was walking up the hill from Zollikon station after leaving your house, I did not actually see any "flying saucers," but I *did* see a particularly beautiful large *meteor*. It was moving relatively slowly (this can be explained by factors of perspective) from east to west and then finally exploded, producing an impressively fine firework display. I took it as a spiritual "omen" that our general attitude toward the spiritual problems of our age is in the sense of καιρός, in other words is more a "meaningful" one.

(In ancient Greek, καιρός is the window of opportunity in which something meaningful can be achieved.)

Pauli went so far as to consult with his scientific colleagues as to whether UFOs really existed. Among others, he corresponded with the eminent German electrical engineer, Max Knoll, who was also an adept of Jungian psychology. Knoll was willing to believe that UFOs resulted from the individuation process but he was adamant that they did not physically exist. He replied at length to queries from Pauli on the nature of radar systems and dismissed every known sighting of UFOs, whether on radar or with the naked eye: "Jung must be made to understand that the UFOs seen on radar screens are no more 'real' than those sighted directly," he wrote tetchily. "[There] are no reliable sightings."

Pauli forwarded Knoll's letter to Jung. Jung's personal secretary, Aniela Jaffé, responded on his behalf: "Professor Jung was much interested in Knoll's letter. But that did not stop him from saying, in a tone of resignation: People think I am more stupid than I am!"

By this time Jung was a venerable figure, receiving scores of visitors at his mansion in Küsnacht, outside Zürich. More than his psychological work, most wanted to hear his views on UFOs. In 1958 he published *Flying Saucers: A Modern Myth of Things Seen in the Skies*. It was greeted with an acclaim that astonished even him.

Among the visitors to Küsnacht who tried to convince Jung that UFOs did not exist was Charles Lindbergh, the first man to fly across the Atlantic. The United States Air Force had investigated hundreds of sightings of UFOs, he told Jung, and found not the slightest evidence to support them. The author of one of Jung's favorite books on flying saucers, Donald Keyhoe, suffered from poor mental health, he added, and the Pentagon had disproved all his allegations.

The commander of the American Air Force, General Carl Spaatz, had told Lindbergh that if UFOs actually existed, surely both of them would have heard about it. Jung snapped, "There are a great many things going on around this earth that you and General Spaatz don't know anything about."

Pauli's last letter to Jung was in August 1957. He continued to send him offprints of his papers. Aniela Jaffé apologized for replying on Jung's behalf, saying that he was very tired. He was now eighty-two. She sent Pauli her "best wishes for 1958—many long journeys."

Carl Jung was to outlive Pauli. Jung died on June 6, 1961, at the age of eighty-five.

15

The Mysterious Number 137

The fine structure constant

PAULI once said that if the Lord allowed him to ask anything he wanted, his first question would be "Why 1/137?"

One of his colleagues mischievously filled in the rest of the story. He imagined that one day Pauli did get the chance to ask his question. In response the Lord picked up a piece of chalk, went to a blackboard and started explaining exactly why the fine structure constant had to be 1/137. Pauli listened for a while, then shook his head. "No," said the man who was famous for declaring of a theory, "Why, that's not even wrong!" He then pointed out to the Lord the mistake that He had made.

The fine structure constant is one of those numbers at the very root of the universe and of all matter. If it were different, nothing would be as it is. As Max Born put it, it "has the most fundamental consequences for the structure of matter in general." To recap: spectral lines are the lines that are the fingerprints of an atom, revealed when they are illuminated by light. The fine structure is the structure of individual spectral lines.

The fine structure constant, in turn, is the immutable figure that defines the fine structure.

Pauli's mentor, Arnold Sommerfeld, calculated the fine structure constant as 0.00729. Later scientists, however, discovered that this could be written as the simpler, more meaningful number 1/137 (that is, 1 divided by 137). It soon became known more familiarly as 137.

This number was beyond discussion. It simply had to be 1/137 because this determines the spacing between the fine structure of spectral lines, as had been discovered in the laboratory. But why this figure? Why 137? There was something about 137—both a prime and a primal number— that tickled everyone's imagination.*

To reiterate: The three fundamental constants that make up the fine structure constant are the charge of the electron, the speed of light, and Planck's constant, which determines the smallest possible measurement in the world. All these have dimensions. The charge of the electron is 1.61×10^{-19} Coulombs, the speed of light is 3×10^8 meters per second, and Planck's constant is 6.63×10^{-34} Joule-seconds. All three depend on the units in which they are measured. Thus the speed of light is 3×10^8 meters per second in the metric system but 186,000 miles per second in the imperial system. All three would certainly play an essential part in a relativity or quantum theory formulated by physicists on another planet in another galaxy, but these physicists might have a different system of measurements from ours and therefore the exact figure would almost certainly not be the same.

The fine structure constant is entirely different. Even though it is made up of these three fundamental constants, it is simply a number, because the dimensions of the charge of the electron, Planck's constant, and the speed of light cancel out. This means that in any number system it will always be the same, like pi which is always 3.141592. . . . So why

---

*The most recent determination of the fine structure constant was a marvel of modern technology. Gerald Gabrielse and co-workers at Harvard University created an artificial hydrogen atom by catching a single electron in a trap made of a positively charged electrode, supplemented by coils producing a magnetic field. They carefully measured the light produced by the electron as it jumps between energy levels and obtained the most accurate value of the fine structure constant to date: $1/\alpha = 137.035999084\ldots$ (Hanneke [2008]).

is the fine structure constant 137? Physicists could only conclude that it cannot have this value by accident. It is "out there," independent of the structure of our minds.

Never before in the history of modern science had a pure number with no dimensions been found to play such a pivotal role. People began referring to it as a "mystical number." "The language of the spectra"—the spectral lines, where Sommerfeld had found it—"is a true music of the spheres within the atom," he wrote.

Arthur Eddington and his mania for 137

In 1957, when Pauli was fifty-seven, he wrote to his sister Hertha:

> I do not believe in the possible future of mysticism in the old form. However, I do believe that the natural sciences will out of themselves bring forth a counter pole in their adherents, which connects with the old mystic elements.

Perhaps the clue lay in numbers—more specifically, the number 137.

Until 1929, the fine structure constant was always written 0.00729. That year the English astrophysicist Arthur Eddington had a bright idea. He tried dividing 1 by 0.00729. The result was 137.17, to two decimal place accuracy. The actual measurement of the fine structure constant as ascertained in the laboratory, Eddington pointed out, was close to that (it was somewhere between 1/137.1 and 1/137.3). There were two ways in which scientists determined the fine structure constant. When they calculated it from the measured values of the charge of the electron, Planck's constant, and the velocity of light, the result was 0.00729. Or they could measure the actual fine structure of spectral lines—that is, determine it in the laboratory; the most frequent value that resulted was $0.007295 \pm .000005$. However, the latter method required input from a particular theory and this presented a problem, for the theory of how electrons interacted with light—quantum electrodynamics—was still in flux, as indicated by the difficulties Heisenberg and Pauli experienced in their work on this very subject. With this in mind, Eddington felt justified in throwing numerical accuracy to the wind and writing the fine structure constant simply as 1/137—the ratio of two prime numbers.

Eddington had a strong mystical streak. To him, mysticism offered an escape from the closed logical system of physics. "It is reasonable to inquire whether in the mystical illusions of man there is not a reflection of an underlying reality," he mused. Like Pauli, he struggled with the dichotomy between the two worlds, both equally invisible, of science and the spirit. He was sure that mathematics was the key that would open the door between these two worlds and he set about an obsessive quest to derive 137 however he could.

The equation for the fine structure constant is

$$\alpha = \frac{2\pi e^2}{hc}$$

and the charge of the electron, e, appears in this equation as $e \times e$, or e^2. As a result, besides being a measure of the fine structure of spectral lines, the fine structure constant (1/137) also measures how strongly two electrons interact.

Eddington argued that according to relativity theory, particles cannot be considered in isolation but only in relation to each other and therefore any theory of the electron has to deal with at least two electrons. Applying a special mathematics that he had invented, Eddington found that each electron could be described using sixteen E-numbers (E stood for "Eddington"). Multiplying 16 by 16 gave a total of 256 different ways in which electrons could combine with each other. He then showed that, of these 256 ways, only 136 are actually possible; 120 are not. He wrote this mathematically as $256 = 136 + 120$. Like pulling a rabbit out of a hat, he thus magically produced the number 136 from purely mathematical (if dubious) reasoning.

Of course 136 was not 137, but for Eddington it was close enough. He was convinced that the elusive "one" would "not be long in turning up." As the physicist Paul Dirac put it, "[Eddington] first proved for 136 and when experiment raised to 137, he gave proof of that!" The obsessive pursuit of 137 took over Eddington's life. American astrophysicist Henry Norris Russell remembered meeting him at a conference in Stockholm.

They were in the cloakroom, about to hang up their coats. Eddington insisted on hanging his hat on peg 137.

Eddington added to the mystery by pointing out that 137 contained three of the seven fundamental constants of nature (the other four are the masses of the electron and the proton, Newton's gravitational constant, and the cosmological constant of the general theory of relativity). Seven, of course, is a mysterious number in itself, encompassing the seven days of creation, seven orifices in the head, and seven planets in the pre-Copernican planetary system. Eddington's speculations were a catalyst in the search for numerical relationships among the fundamental constants of nature.

In January 1929 Bohr wrote to Pauli, "What do you think of Eddington's latest article (136)?" Pauli hardly bothered to reply, commenting only that he might soon have something to say "on Eddington (??)." The two question marks are his. "I consider Eddington's '136-work' as complete nonsense: more exactly for romantic poets and not for physicists," he wrote to his colleague Oskar Klein a month later. He added in a letter to Sommerfeld that May, "Regarding Eddington's $\alpha = 1/136$, I believe it makes no sense." (α is the fine structure constant.)

Pauli and 137

It seemed that Pauli had not caught the 137 bug. In February 1934, however, he wrote to Heisenberg that the key problem was "fixing [1/137] and the '*Atomistik*' of the electric charge." At the time he was trying to find a version of quantum electrodynamics in which the mass and charge of the electron were not infinite; but no matter which way he manipulated his equations, the concept of electric charge always entered—hence the mystical " '*Atomistik*'—atom plus mystic—of the electric charge."

The problem was that quantum electrodynamics did "not take the atomic nature of the electric charge into account" when the electric charge entered the theory of quantum electrodynamics as part of the fine structure constant (that is, 1/137). "A future theory," Pauli wrote, "must bring about a deep unification of foundations."

As Pauli saw it, the crux of the problem was that the concept of electric charge was foreign to both prequantum and quantum physics. In both theories the charge of the electron had to be introduced into equations—it did not emerge from them. (This was similar to Heisenberg and Schrödinger's quantum theories in which the spin of the electron had to be inserted, whereas it popped out of Dirac's theory.)

Quantum theory exacerbated this situation in that it included the fine structure constant, $1/137 = 2\pi e^2/hc$, that is, it linked the charge of the electron (e) with two other fundamental constants of nature—the miniscule Planck's constant, h (the smallest measurement possible in the universe and the signature of quantum theory which deals with nature at the atomic level), and the vast speed of light, c (the signature of relativity theory which deals with the universe).

Pauli continued to worry about the connection between the fine structure constant and the infinities occurring in quantum theory. It was a problem that would not go away. "Everything will become beautiful when [1/137] is fixed," he wrote to Heisenberg in April 1934. And on into June: "I have been musing over the great question, what is [1/137]?"

That year, in a lecture he gave in Zürich, he underlined the importance of eliminating the infinities that persisted in quantum electrodynamics and drew attention to the theory's relationship to our understanding of space and time. The solution to this problem would require "an interpretation of the numerical value of the dimensionless number [137]."

So what had happened? Why did Pauli suddenly begin to discuss his thoughts on 137? Perhaps it was the effect of Jung's analysis opening his mind to mystical speculations.

In 1935, the senior scientist Max Born, Pauli's mentor at Göttingen who was then at Cambridge, published an article entitled, "The Mysterious Number 137." He looked into the reasons why 137 should have such mystical power for scientists. The main reason was that it seemed to be a way in which one could achieve the Holy Grail of scientific studies—linking relativity (the study of the very large—the universe) with quantum theory (the study of the very small—the atom).

In his article he looks at some of the qualities that make the number "mystical," prime among them being that even though it is made up of fundamental constants that possess dimensions, it is itself dimensionless.

It is also enormously important in the development of the universe as we know it. He writes: "If α [the fine structure constant] were bigger than it really is, we should not be able to distinguish matter from ether [the vacuum, nothingness], and our task to disentangle the natural laws would be hopelessly difficult. The fact however that α has just its value 1/137 is certainly no chance but itself a law of nature. It is clear that the explanation of this number must be the central problem of natural philosophy."

In 1955, at the hundredth anniversary of the ETH, Pauli addressed a huge audience in the main lecture hall of the physics department at Gloriastrasse on the subject "Problems of Today's Physics." Contrary to his usual style Pauli spoke from a prepared manuscript. He obviously found this difficult. With a flourish he threw the paper aside and spoke off the cuff with great passion and verve. The crux of his argument was the vital importance of the fine structure constant and also what an impenetrable problem it was. It did not merely designate how two electrons interact with each other, it was not merely a constant to be measured; what scientists had to do was "to accept it as one of the actual main problems of theoretical physics." There was thunderous applause.

Realizing its fundamental importance in understanding spectral lines in atomic physics and in the theory of how light and electrons interact, quantum electrodynamics, Pauli and Heisenberg were determined to derive it from quantum theory rather than introducing it from the start. They believed that if they could find a version of quantum electrodynamics capable of producing the fine structure constant, it would not contain the infinities that marred their theories. But nothing worked. The deeper problems that beset physics—not only how to derive the fine structure constant but how to find an explanation for the masses of elementary particles—remain unsolved to this day.

In 1985 the brash, straight-talking, American physicist Richard Feynman, who had studied Eddington's philosophical and scientific papers on 137, wrote in his inimitable manner:

It [1/137] has been a mystery ever since it was discovered more than fifty years ago, and all good theoretical physicists put this number up on their wall and worry about it. Immediately you would like to know where this number comes from. . . . Nobody knows. It's one of the *greatest* damn mysteries of physics: a *magic number* that comes to us with no understanding by man.

The magic number 137

So where did this magic number come from? How did Sommerfeld, who discovered it, alight on it? To get a glimpse of his thought processes, we need to take a short mathematical journey.

Mulling over the problem of the structure of spectral lines, Sommerfeld took another look at the key equation in Bohr's theory of the atom as a miniscule solar system. It is

$$E_n = -\frac{Z^2}{n^2} \ (2.7 \times 10^{-11} \ \text{ergs})$$

This is an equation for the energy level of the lone electron in an atom's outermost shell, such as the electron in a hydrogen atom or in alkali atoms—hydrogen-like in structure, in that they have one electron free for chemical reactions, while all the rest are in closed shells (Pauli studied them in his work on the exclusion principle); ergs are the units in which energy is expressed. The equation shows the energy (E) of the electron in a particular orbit designated by the whole quantum number n. Z is the number of protons in the nucleus; and the minus sign indicates that the electron is bound within the atom. The quantity 2.7×10^{-11} ergs results from the way in which the charge of the electron (e), its mass (m), and Planck's constant (h) occur in this equation:

$$2.7 \times 10^{-11} \ \text{ergs} = \frac{2\pi^2 m e^4}{h^2}$$

It is also the energy of an electron in the lowest orbit ($n = 1$) of the hydrogen atom ($Z = 1$).

Sommerfeld decided that the mathematics in Bohr's original theory needed to be tidied up. His brilliant idea was to include relativity in the new mathematical formulation, making the mass of the electron behave according to $E = mc^2$, in which mass and energy are equivalent. This was the result:

$$E_{n,k} = -\frac{Z^2}{n^2}\left\{1 + \left(\frac{2\pi e^2}{hc}\right)^2\left[\frac{n}{k} - \frac{3}{4}\right]\right\} (2.7 \times 10^{-11} \text{ ergs}) \,.$$

In this new equation, the additional quantum number k indicated the additional possible orbits for the electron and allowed the possibility for an electron to make additional quantum jumps from orbit to orbit. It therefore also allowed the possibility of the atom having additional spectral lines—a fine structure.

The first term in the equation for $E_{n,k}$—outside the brackets—was the same as in Bohr's original equation. But a whole extra term had appeared inside the large brackets.

Multiplying this term was an extraordinary bundle of symbols that no one had ever seen before: $\left(\frac{2\pi e^2}{hc}\right)$. In this expression, e is the charge of the electron, h is Planck's constant, and c is the velocity of light. Sommerfeld deduced the number 0.00729 from this bundle of symbols. He realized that this was the number that set the scale of the splitting of spectral lines—that is, of the atom's fine structure—and called it the fine structure constant. It is there in the equation because it is there in the atom; it is part of the atom's existence, which includes the fine structure of a spectral line. Physicists knew the fine structure existed. They had measured the fine structure splitting, but they didn't have an equation for it that agreed with experiment. Now they did:

$$E_{n,k} = -\frac{Z^2}{n^2}\left\{1 + \left(\frac{1}{137}\right)^2\left[\frac{n}{k} - \frac{3}{4}\right]\right\} (2.7 \times 10^{-11} \text{ ergs})$$

This extraordinary equation, in which $2\pi e^2/hc$ is replaced by $1/137$, perfectly described the fine structure of spectral lines as observed in experiments.

Not just scientists but many others have grappled with 137. For a start, 137 can be expressed in terms of pi. Some complicated ways of doing this, all of which end up with 137, are

$$1/\alpha = 8\pi(8\pi^5/15)^{1/3} = 137.348$$
$$1/\alpha = (8\pi^4/9)(2^4 5!/\pi^5)^{1/4} = 137.036082$$

We can also write 137 as a series of Lucas numbers, which are connected with Fibonacci numbers and the Golden Ratio.

Fibonacci numbers is a sequence of whole numbers in which each number starting from the third is the sum of the two previous ones. The sequence begins 0, 1, 1, 2, 3, 5, 8, 13, 21, 34, 55, 89, 144, 233, 377, 610, 987, and so on (0 + 1 = 1, 1 + 1 = 2, etc.).

If we take the ratio of successive numbers in the series—1/1 = 1.000000, 2/1 = 2.000000, 3/2 = 1.500000, 5/3 = 1.666666, and so on to 987/610 = 1.618033—we reach 1.6180339837, the Golden Ratio, which has been a guideline for architecture since the days of ancient Greece. It appears on the pyramids of ancient Egypt, the Parthenon in Athens, and the United Nations building in New York City.

Fibonacci numbers were discovered by the Italian mathematician Leonardo Fibonacci in the twelfth century. It was Kepler who discovered their relation to the Golden Ratio. Then Edouard Lucas, a French mathematician, used them to develop the Lucas numbers in the nineteenth century.

Lucas numbers are like Fibonacci numbers but begin with 2: 2, 1, 3, 4, 7, 11, 18, 29, 47, 76, 123, and so on. Like the Fibonacci series, the Lucas series also produces the Golden Ratio.

Fibonacci and Lucas numbers pop up all over the place, from how rabbits reproduce to the shape of mollusk shells, to leaf arrangements that sometimes spiral at angles derivable from the Golden Ratio.

Because all these numbers are related, any formula for 137 in terms of the Golden Ratio can be rewritten in terms of Fibonacci and Lucas numbers, though whether this is anything more than merely abstruse relationships between certain numbers is not clear.

If one tries hard enough, 137 can be deduced from devilishly complicated combinations of "magic numbers" such as 22 (the first 22 human chromosomes are numbered), 23 (the number of chromosomes from each parent), 28 (the length of a woman's menstrual cycle), 46 (the number of pairs of chromosomes a person has), 64 (the number of possible values for the 20 amino acids in DNA), and 92 (the number of naturally occurring elements in the periodic table).

Sadly, all these are pure coincidences with no scientific basis. And still 137 continues to tantalize. In fact 137 has become something of a cult.

Enrico Fermi with his incorrect equation for the fine structure constant.

According to one Web site, "The Fine Structure Constant holds a special place among cult numbers. Unlike its more mundane cousins, 17 and 666, the Fine Structure Constant seduces otherwise sane engineers and scientists into seeking mystical truths and developing farfetched theories."

Even Heisenberg had a go, fired by Eddington's number speculations. In a letter to Bohr in 1935 he reported "playing around" with the fine structure constant, which he expressed as $\frac{\pi}{2^4 3^3}$. He was quick to add, "but the other research on it is more serious," referring to his and Pauli's attempts to derive it from quantum electrodynamics.

Many years later, Enrico Fermi, the physicist who christened Pauli's newfound weakly interacting particle the neutrino, was asked to pose for a photograph. He took his place in front of a blackboard on which he had written the fine structure constant—but incorrectly. Instead of $\left(\frac{2\pi e^2}{hc}\right)$ he put $\left(\frac{\hbar^2}{ec}\right)$. ($\hbar$ is shorthand for Planck's constant h divided by 2π: $h/2\pi$.)

It was an excellent joke—and a joke comprehensible only to scientists. However, the joke backfired when the photograph was used for a stamp to commemorate him after his death. So he is caught forever standing next to this iconic equation—incorrectly written.

Surprisingly, 137 also crops up in an entirely different context. In the 1950s, Pauli developed a close friendship with Gershom Scholem, a prominent scholar of Jewish mysticism. When his former assistant Victor Weisskopf went to Jerusalem, Pauli urged him to meet Scholem, though he gave no hint of his own interests in Jewish mysticism. Scholem asked Weisskopf what the deep unsolved problems of physics were. Weisskopf replied, "Well, there's this number, 137."

Scholem's eyes lit up. "Did you know that 137 is the number associated with the Kabbalah?" he asked.

In ancient Hebrew, numbers were written with letters, and each letter of the Hebrew alphabet has a number associated with it. Adepts of the philosophical system known as the Gematria add the numbers in Hebrew words and thus find hidden meanings in them. The word Kabbalah is written קבלה in Hebrew: ה is 5, ל is 30, ב is 2, and ק is 100. The four letters add up to . . . 137!

It is an extraordinary link between mysticism and physics. Two key words in the Kabbalah are "wisdom," which has a numerical value of 73, and "prophesy" (64): 73 + 64 = 137. God himself is One—1—which can also be written 10 (1 + 0 = 10). Take 10's constituent prime numbers, 3 and 7, and add the original 1: together they can be written 137.

In the bible the key phrase "The God of Truth" (Isaiah 65) יהלא זמא adds up to 137. So does "The Surrounding Brightness" (Ezekiel 1) סביב הנגה, and the Hebrew word for "crucifix", צליבה.

It turns out that, according to the Gematria, the number content of the letters in the Hebrew word for 137 add up to 1664. This happens to be the numerical value for a portion of the well-known passage from Revelations 13:18: "Here is wisdom. Let him that hath understanding count the number of the beast." The rest of the passage reads: "for the number is that of man; and his number is six hundred and sixty-six."

Homing in on the yet more arcane, a group of numerologists noticed that the number 8294 appeared at key places in Aztec and Roman texts and went on to argue that it was a key to a "new universal consciousness."

Their confidence in this assertion was bolstered when they discovered they could relate it to the fine structure constant and the number of the beast—666—as follows:

$$\frac{10^{82943.99629/(32\times666)}}{666} = \sqrt{137}$$

And more way out still, 137 is the foundation stone of a fiendishly complex "biblical mathematics" referred to by followers of "The Bible Wheel" as a "Holographic Generating Set." It is based on three geometric forms: a cube, A, divided into 27 subcubes; a hexagon, B, divided into 37 subhexagons; and a star of David, C, divided into 73 circles Of course $27 + 37 + 73 = 137$. From this "generating set" with its Pythagorean-geometric aura, aficionados of the Bible Wheel claim to be able to generate biblical passages and probe mystical numbers by multiplying A, B, and C in various ways.

Thus 137 continues to fire the imagination of everyone from scientists and mystics to occultists and people from the far-flung edges of society.

The last challenge

With World War II behind them, Heisenberg and Pauli resumed their scientific correspondence. But the days of their collaboration and the frequent exchange of letters seemed to have ended. After all, they had been on opposite sides in the war. And Heisenberg's reputation was colored by the fact that he had remained in Germany and had ended up in charge of the German atomic bomb project. Most of his postwar colleagues, of course, had been involved in the Manhattan Project. Even Heisenberg's brilliance was not enough to overcome this stain.

Then in 1957 Heisenberg wrote to Pauli that he had the germ of a theory that could explain the masses of elementary particles as well as most of the symmetries. In a preliminary test he was almost able to deduce the fine structure constant from his new theory. In his calculations it came to 1/250, which is not very far away from 1/137, in the same way as 1/3 is not far from 1/4, even though 3 is far from 4. This was extraordinary, given that Heisenberg's theory was still in its formative stages. Pauli immediately took up Heisenberg's suggestion to join him in his project.

"Never before or afterward have I seen [Pauli] so excited about physics," Heisenberg later recalled. It seemed as if the old days had come back. The two giants of quantum physics were working together once again.

"The picture keeps shifting all the time. Everything in flux. Nothing for publication yet but it's all bound to turn out magnificently," Pauli wrote exuberantly to Heisenberg at the start of 1958. "This is powerful stuff. . . . The cat is out of the bag and has shown its claws: division of symmetry reduction. I have gone out to meet it with my antisymmetry— I gave it fair play—whereupon it made its quietus. . . . A very happy New Year. Let us march forward toward it. It's a long way to Tipperary, it's a long way to go."

Heisenberg was equally excited. For him the theory was to be the culmination of his life's work. Over the years he had become entranced by the power of mathematics to probe and understand the physical world. And he knew how to use it. "A wonderful combination of profound intuition and formal virtuosity inspired Heisenberg to conceptions of striking brilliance," a colleague wrote. He had used his formidable insight and daring to apply mathematics to make his startling discoveries in quantum mechanics. These included the uncertainty principle; the first steps toward understanding the force that holds the nucleus together; and his attempts to produce a coherent theory of electrons and light, known as quantum electrodynamics.

The theory he was working on with Pauli was exactly what he was looking for. "The last few weeks have been full of excitement for me," he wrote to his sister Edith in January 1958:

> I have attempted an as yet-unknown-ascent to the fundamental peak of atomic theory with great efforts during the last five years. And now, with the peak directly ahead of me, the whole terrain of interrelationships in atomic theory is suddenly and clearly spread out before my eyes. That these interrelationships display, in all their mathematical abstraction, an incredible degree of simplicity, is a gift we can only accept humbly. Not even Plato could have believed them to be

so beautiful. For these interrelationships cannot have been invented; they have been there since the creation of the world.

Perhaps he also saw it as his chance to vindicate himself, to put behind him the fact that he had been the leader of the German atomic bomb project.

Together Pauli and Heisenberg wrote a joint paper that Pauli planned to lecture on during his forthcoming visit to the United States that January. Before leaving he wrote to Aniela Jaffé. He had been entirely engrossed in work with Heisenberg on a "new physical-mathematical theory of the smallest particles," he said.

The pace of their work was so overwhelming that often letters were not fast enough. The phone line between Zürich and Munich, where Heisenberg was based, was continually buzzing. For Pauli their research was so fundamental that he saw Jungian significance in it. Although he and Heisenberg were very different, he wrote to Jaffé, they were "gripped by the same archetype"—by reflection symmetry, in the fullest sense of the *CPT* reflection. "Director *Spiegler*! [Reflector] dictates to me what I should write and calculate," he declared.

He hoped that the "new year will see a beautiful theory that will light up the world." He had even had a dream about it: Pauli enters a room and finds a boy and a girl there. He calls out, "Franca, here are two children!" He had seen Heisenberg just three days earlier and interpreted the children as the new ideas which he was confident would emerge from their work. From the fact that there were two children, he drew an analogy to his "mirror complex."

The work could be taken as a realization of the unconscious and "more specifically: a *realization* of the 'Self' (in the Jungian sense)." It was what Pauli had sought for years—physics and the psychology of the unconscious as mirror images, a scenario that had been destroyed by the violation of mirror symmetry (parity violation), but restored by *CPT* symmetry, Pauli's 1955 discovery about which he was still exultant.

On February 1, 1958, the lecture theater at the physics department at Columbia University was packed with over three hundred people. They were eager to see the great Pauli who was about to lecture on a theory formulated by the two giants of quantum theory. Niels Bohr,

J. Robert Oppenheimer, T. D. Lee, and C. N. Yang, who had proposed the overthrow of parity, and C. S. Wu, the physicist who had performed the crucial experiment to prove it, all attended. The air was electric. But the distinguished audience had nothing but criticism for the new theory, though offered in a friendly manner. A key point in the theory was how newly discovered elementary particles decayed, how they transformed themselves into other particles. As Pauli was scribbling the equations on the blackboard, Abraham Pais, an eminent physicist and friend of Pauli's, raised his hand and objected, "But Professor Pauli, this particle does not decay like that." Pauli stopped midflow. There was a long silence. Then Pauli muttered, "I must get in touch with my friends in Göttingen about that," by which he meant Heisenberg. At this point, T. D. Lee remembers, you "could almost feel the silence." Others too pointed out loopholes in his mathematical proofs. Pauli continued his lecture but it was clear that the passion had gone.

At one point Bohr and Pauli chased each other around a long table at the front of the room. Whenever Bohr ended up at the front he declared, "It is not crazy enough." Each time Pauli appeared he replied, "It is crazy enough." This was repeated several times and the audience burst into applause.

"We were all polite, but Pauli was obviously discouraged . . . It was obvious that his heart was no longer in their work," Yang recalled. He vividly remembered Pauli's gloom as they were on their way in Lee's car to a restaurant after the lecture. "Pauli oscillated back and forth in his seat and murmured some thing which I thought was, 'As I talked more and more, I believed in it less and less.' I was greatly saddened." The physicist Freeman J. Dyson, of the Institute of Advanced Study, commented that it was "like watching the death of a noble animal."

So what had happened? How could Pauli have been so enthusiastic about this new approach and then so totally cast down? Presenting a lecture on a subject can shed an entirely new light on it. So perhaps that was what happened. Suddenly he realized that the theory was full of holes. He was beginning to have second thoughts about his work with Heisenberg.

The following day he lectured at the American Physical Society meeting in New York, at that time the biggest and most important annual gathering of physicists from around the world. There was standing room only. But the criticism meted out by a younger and brasher generation of American physicists was even harsher. Lee could not bring himself to attend.

From there Pauli went on to California—where Feynman, for one, had had no compunction about telling the great Bohr that he was an idiot. Audiences there too offered ruthless criticism. Pauli was beginning to conclude, as he wrote to Heisenberg later that year, that "something entirely new, in other words very 'crazy,' [was] needed" if he and Heisenberg were to crack the mystery of the masses of elementary particles, one of the key aims of their unified theory.

Disillusioned, Pauli attacked Heisenberg's calculation of the fine structure constant as 1/250, which had seemed so promising and had played a part in his decision to join Heisenberg's project. He wrote to Fierz bitterly, "I have never considered it as correct. It's so totally stupid."

Some time later, Heisenberg's co-worker on that calculation, Roberto Ascoli, recalled that he had originally deduced the fine structure constant as 8 on the basis of Heisenberg's theory. "Only after Heisenberg had doctored it up, was the value reduced to 1/250."

Later that month Heisenberg gave a lecture on his and Pauli's work at Heisenberg's Institute in Göttingen. The room was packed. The great Heisenberg was about to announce a theory that could explain the behavior of every elementary particle in the world with a single equation—a "world formula"—that would surely prove to be an abstruse and highly technical piece of mathematics.

A press release was circulated reading (most offensively to Pauli): "Professor Heisenberg and his assistant, W. Pauli, have discovered the basic equation of the cosmos." The story was picked up by newspapers around the world. Pauli vented his anger in a letter to George Gamow, the physicist and prankster who had translated and illustrated the Mephistopheles spoof at Bohr's Institute in 1932. Pauli lampooned it by drawing an empty box saying, "This is to show that I can paint like Titian: Only the technical details are missing"—a case of the emperor's new clothes. Pauli requested Gamow not to publish his comment but, miffed at Heisenberg's insulting press release, added, "please show it to other physicists and make it popular among them." Gamow certainly did. A week later

Weisskopf wrote to Pauli about the press release and added that he read it with Pauli's comment about Titian well in mind.

Combining Pauli's well-known obsession with 137 with Heisenberg's for his new theory, a colleague wrote jokingly to Pauli, "Since the Heisenberg equation is supposed to describe everything (see, for instance, *New-York Herald Tribune*, volume 137, p. i/137), it has as one of its solutions Heisenberg himself." Pauli replied, "Regarding Heisenberg I have the feeling, that the situation is slowly growing over his head; certainly he needs vacations." The problem of how to describe the properties of elementary particles remained open. Pauli summed up the situation thus, "Many questions, no good answers."

In fact, Pauli was furious. He wrote to C. S. Wu about Heisenberg's "poor taste" as far as the press releases were concerned:

> In some of these I had been, unfortunately, mentioned . . . but fortunately only in a "mild" form as a secondary (or tertiary) auxiliary person of the Super-Faust, Super-Einstein and Super-man Heisenberg. (He seems to have mentioned his dreams on gravitational fields—about which one has *not* worked at all in Göttingen recently—and his revival of the old idea of a "world formula"—which was never successful—in a quantized form.)

He seems to have been relieved that he was not associated too closely with Heisenberg's mistaken schemes. To Wu he recalled his hopes, dreams, and aspirations of thirty years earlier when he and Heisenberg were young and Heisenberg depended on his friend's criticism and inspiration. Now Pauli had had enough:

> Heisenberg's desire for publicity and "glory" seems to be insatiable, while I am in this respect completely saturated. I only need something in science which interests me sufficiently and with which I can play (without being a hero in the limelight of the "world.") Heisenberg's opposite attitude, with which he certainly wishes to compensate earlier failures, may have many reasons lying in the whole history of his life.

No doubt the last words were an oblique reference to the role Heisenberg played in the war.

Soon afterward Pauli withdrew from the collaboration. Heisenberg persisted in making promise after promise as to the wonders his theory would produce. "He believes that if he publishes with me, then it is 1930 again! I have found it embarrassing how he runs after me!" Pauli wrote to Fierz in May.

That July Pauli chaired a session at a conference at CERN at which Heisenberg was scheduled to speak. Pauli introduced Heisenberg with the words, "What you will hear today is only a substitute for fundamental ideas." He went on to make a request of the audience: "don't laugh in the wrong place, ha, ha, ha. . . ." The audience was already in fits of laughter. Pauli let Heisenberg finish speaking, then mercilessly demolished his paper.

When Pauli and Heisenberg met again later that summer, Heisenberg noticed that Pauli looked dispirited. Pauli encouraged him to go on with his work and wished him well, but added, "For me, I have to drop out, I just haven't the strength, and that's that. Things have changed too much."

Could it have been that the great criticizer had met his match in the lambasting he encountered in America? To be on the receiving end must have been shattering. Heisenberg had been afraid this would happen when Pauli "in his present mood of exultation [encountered] the sober American pragmatists." Franca too had noticed this chink in Pauli's armor which, up until then, he had concealed so successfully: "He was very easily hurt and therefore would let down a curtain. He tried to live without admitting reality. And his unworldliness stemmed precisely from his belief that that was possible."

But there was something else that brought Pauli to the point of spiritual exhaustion. He had grown attached to his work with Heisenberg. Their new theory had all the trappings that he thought a theory should have: a high degree of mathematical symmetry and Jungian meaning too, taking it one step closer to not merely a unified theory of elementary particles from which the fine structure constant could one day be deduced, but to a theory of the mind as well.

As he always did, Pauli interpreted the failure of the theory as personal failure. Genius though he was, he had failed yet again. Those who saw him in the autumn of 1958 recalled that he seemed beaten.

That year Pauli told an interviewer, "When I was young I thought I was the best formalist of my time. I believed that I was a revolutionary. When the great problems would come, I would be the one to solve them and to write about them. Others solved them and wrote about them. I was but a classicist and no revolutionary." He began writing letters as if saying farewell.

A different side of Pauli

It was not all gloom. During his visit to the United States that year, Pauli had visited Harvard. Among the delegation that greeted him was Roy Glauber.

Glauber had been a postdoctoral fellow at the ETH back in 1950, working under Pauli. "Pauli . . . had been a legendary figure since his early twenties. Some part of that legend, as Pauli well knew, was attached to his role as a critic, not always kindly, of the work of his colleagues. So no one who knew Pauli, it is fair to say, could be in his presence without feeling a certain defensive wariness," he remembered many years later.

Not long after young Glauber arrived in Zürich, he and the rest of Pauli's students went on a hike in the hills. They took a cable car and then followed a series of steep trails around the Vierwaldstätter See, a scenic lake. "Pauli, notwithstanding his ample girth," kept up a vigorous pace. Expecting a picnic, Glauber had brought a camera, a hefty Speed Graphic—made famous in the 1940s and 1950s by press photographers—which hung by a strap over his shoulder. He had not even brought much film. Pauli began teasing him. "Always you carry that awful camera," he kept saying, "but you are taking no pictures." Then he laughed uproariously.

At the end of the day, some of the group swam in the lake while others played soccer. Pauli kicked the ball into the lake so that someone had to swim out and fetch it. He roared with laughter as he kicked the ball further and further into the water.

A photograph of Pauli kicking the ball was too good to miss—but Glauber had only one exposure left. Trying not to draw Pauli's attention,

Pauli kicking a soccer ball, 1950.

he set up his camera, peered into the rangefinder and as inconspicuously as possible signaled a friend to kick the ball toward Pauli. Suddenly the camera smacked him square in the face. Instead of the lake, Pauli had decided to make Glauber's camera his next target. Glauber recalled hearing his bellowing laugh.

He had, however, managed to snap the shutter. "Pauli," he concluded, "never discounted the element of luck in his practical jokes."

When Glauber was still at the ETH, his mother once wrote and complained to Pauli that her son never sent letters home. Thereafter, whenever Pauli saw Glauber, he always insisted on asking loudly, "And how is your dear mother?"

In 1958, when Pauli visited Harvard, Glauber was apprehensive. To his relief Pauli greeted him warmly and said nothing the entire time about his mother. "Thank God he has forgotten my mother," Glauber said after he had left. Pauli went with Weisskopf back to his lodging in Cambridge. The first thing he said once they were out of earshot was, "This time I fooled Glauber. I said nothing about his mother."

A nearly perfect sphere

That same year Pauli attended the Solvay Conference in Brussels as that venerable meeting's vice president. There he and Franca invited the cosmologist Fred Hoyle and his wife to lunch. Honored, Hoyle happily accepted. He was eager to substantiate a story about Pauli that had been on his mind for some years, of how, in the 1920s, after a lecture by Einstein on relativity, Pauli had had the temerity to say to the audience that what Professor Einstein said "was not so stupid."

But Pauli had his own agenda. "Aha," he cackled, "I just read your novel *The Black Cloud*. I thought it much better than your astronomical work." Hoyle had recently proposed a theory asserting that the universe around us has always existed exactly as we see it. He called it the "steady-state theory" to distinguish it from what he dubbed the "big bang theory"—that is, that the universe came into being at a specific moment in time and evolved into what we see today. Pauli was far from impressed with Hoyle's theory.

He told Hoyle that both he and Jung had read Hoyle's novel carefully and that he was writing a critical essay on it. Hoyle was mystified. After all, it was only a story—about how an intelligent life-form learns to communicate with earthlings. He had never felt it merited such deep analysis.

Finally Hoyle had the chance to ask Pauli about the Einstein story. "My abiding memories of Pauli," he wrote, "are of his helpless laughter as the youthful remark about Einstein came back to him and his rolling back and forth, a nearly perfect sphere." But Pauli said no more, "so I never quite had it from Pauli personally that the story was true, but those who knew him well assure me it was." Another lasting memory that Hoyle carried away from the lunch was the four bottles of fine wine on the table.

The Pauli effect strikes again

That same year the physicist Engelbert Schucking visited Pauli in Zürich. Along with Pauli's assistant Charles Enz and another colleague they took a tram from the ETH to Bellevue Square, where they planned to have a "wet after-session," with plenty of drinking. Bellevue Square is a bustling intersection where several tram tracks cross each other in a seemingly random way. Just as they reached the square, two street cars collided right in front of them with an enormous bang. Schucking was standing with Pauli next to the driver of the street car. "Pauli's face was flushed as he excitedly turned to me and exclaimed, 'Pauli effect!' " Schucking recalled.

Enz told Schucking about a lecture Pauli had given to an audience of high-level government officials on an occasion honoring Einstein, the ETH's most famous graduate. "Pauli read from his manuscript. Whenever he found an error in his text, he stopped in mid-sentence, drew out his fountain pen, corrected the text and went on, oblivious of the squirming audience." It was a teaching style he had maintained throughout his career.

That November Pauli was in Hamburg. Schucking took a walk with him. As they walked along the Gojenbergsweg in the Bergdorf district, looking out over the marshland of the river Elbe, Pauli said several times how glad he was to have withdrawn his name from the paper with Heisenberg.

On Friday, December 5, 1958, as he was teaching his afternoon class, Pauli suddenly began to suffer excruciating stomach pains. Up until then he had been fine. The next day he was rushed to the Red Cross Hospital in Zürich. Charles Enz visited him the day after. Pauli was visibly agitated. Had Enz noticed the number of the room, he asked him?

"No," replied Enz.

"It's 137!" Pauli groaned. "I'm never getting out of here alive."

When the doctors operated, they found a massive pancreatic carcinoma. Pauli died in Room 137 on December 15. His last request had been to speak to Carl Jung.

PAULI was cremated on December 20 and later that afternoon an official funeral ceremony was held at the Fraumünster Church in Zürich, which dates back to the Carolingian period. The ceremony was non-religious. Niels Bohr, Markus Fierz, the party-giver Adolf Guggenbühl,

Pauli's treacherous colleague Paul Scherrer, and his one-time assistant Victor Weisskopf all gave addresses. Franca arranged the funeral. Only physicists spoke. Among the many who attended were Pauli's confidant Paul Rosebaud. Jung, now eighty-two, was relegated to a place at the back. Despite his long association and close friendship with Pauli, he was not invited to speak.

One notable absentee was Heisenberg. The ETH had sent Heisenberg's invitation on the sixteenth, giving him plenty of time to travel from his home in Munich to Zürich to pay his last respects to the man who had been his lifelong friend and colleague and had sparked his greatest discoveries. Heisenberg did not even bother to write a letter of condolence to Franca but left it to his wife. He was, she wrote, reading Pauli's philosophical writings, but due to the Christmas season they were too busy to attend.

It is extraordinary that Heisenberg would spurn his old friend in this way. Despite their recent falling out, one would have assumed that he would have put all that behind him and attended. The only possible explanation is to be found in Heisenberg's autobiography, written over a decade after Pauli demolished the theory he was so proud of. "Wolfgang's attitude to me was almost hostile," he wrote of that episode. "He criticized many details of my analysis, some, I thought quite unreasonably." Presumably Heisenberg never forgot what had happened. The intensity with which these men treated their passion—physics—went far beyond the grave.

Hertha also did not attend her brother's funeral. Perhaps travel was difficult financially for her; perhaps she wanted to remember Wolfgang the way he was; or perhaps she felt uneasy around Franca. Hertha had married E. B. Ashton, an immigrant from Munich, born Ernst Bach. He was a professional translator and they collaborated on several of her books. As she recalled, "we decided he would remain my 'better English' " and married in 1948. As Franca put it contemptuously, she "married her translator." The couple lived happily in a large farmhouse in Huntington, Long Island. Like her brother, Hertha had no children. She published several biographies and historical studies in addition to her books for children on Catholic themes. In the course of her prolific career she became an eminent member of PEN—the worldwide association of writers—and was awarded the Silver Medal of Honor by the Austrian government in 1967. Hertha died in 1973, predeceasing her husband by ten years. Their ashes were interred in the Schütz family grave in Vienna, near where she and Wolfi had grown up.

Pauli's ashes were interred in the graveyard in the town where he had lived, Zollikon, between Zürich, where the ETH is, and Küsnacht, where he used to visit Jung in his Gothic mansion—the two places that defined his two worlds of physics and psychology.

Franca was curious about the story Enz told about room 137. Pauli had never said a word to her about this mysterious number. Enz assured her that he was repeating *"Pauli's own words."* He told her about the significance of the number in physics and that Pauli had mentioned it many times.

Franca wrote a letter to Abdus Salam, a physicist whom Pauli had greatly respected, asking whether it was he who had written "an article connected with the subject Pauli and the number 137." She added, "it is a strange fact that Wolfgang Pauli actually died in the room Nr. 137." People thought that Pauli had requested that room, she said. In fact he had originally been in another room and was transferred to room 137 without being told where he was being sent.

It was Salam, in fact, who originated the story of Pauli going to heaven and asking the Lord to explain "Why 137?" That had been in 1957, the year before Pauli's death. Salam wrote to Franca, "of course it is a story which I would not repeat now." He sent her a copy of his lecture in which the story appeared.

Franca replied, "At last I got a written, beautiful explanation of this to me so elusive Number 137. I enjoyed the end of the story—I did not know—'convincing the Lord a mistake had been made.' One could not characterize Pauli better in so few words!"

FRANCA died in 1987. She spent the three decades after her husband's death finding suitable places for his books, personal papers, and correspondence. She also did her best to delay the publication of his correspondence with Jung. To the end Franca believed that it would detract from his image as a serious scientist.

Epilogue:
The Legacy of Pauli
and Jung

PAULI AND JUNG were men who thought outside the box. Pauli made three discoveries that changed the course of science and our understanding of the world: the exclusion principle, the neutrino, and *CPT* symmetry. Jung pioneered a different way to explore the mind, by opening up psychoanalysis to include alchemy, mysticism, and Far Eastern religions.

Today Pauli is remembered for the exclusion principle and the Pauli legend—as the man who loved to terrorize physicists. He's also remembered, of course, for his extraordinary effect on mechanical instruments. In 2000 the magazine *Physics World* asked scientists to vote for the top-ten physicists of the twentieth century. Pauli did not receive a single vote and was not even mentioned. Yet, beside his three major discoveries, his discussions with and suggestions to Heisenberg (who, of course, was high on the list) were invaluable to Heisenberg in achieving his breakthroughs, as were Pauli's critical evaluations of the work of others. Pauli was involved in some of the greatest advances in twentieth-century physics, but, as we have seen, he couldn't be bothered to step forward and claim the credit. He was more interested in pressing on with his work, in pushing forward the borders of science.

As for Jung, although his name is widely known to psychoanalysts and the general public, what he actually did is less well known than the man himself. For many years the psychoanalytic community spurned him because of his interest in what they saw as the occult. In part this was due to the lambasting he received from Freud's circle. He was later adopted by New Age movements, which did not help matters; nor did his alleged Nazi sympathies. Today there is renewed interest in the connection between Freud and Jung. Scholars are also trying to understand Jung and his work within the culture of his day, both when he was forming his ideas and later as a practitioner. To this end scholars are examining his unpublished papers and preparing them for publication, seeking out further information about who this extraordinary man really was, who started a whole new school of psychology and whose work forms much of the basis of psychology as we have it today.

Today scientists, psychologists, and neurophysiologists routinely assert that the understanding of the mind, including consciousness, cries out for an approach that crosses disciplines. But when Pauli and Jung embarked on this same route it was so innovative—so totally out of the box—that they had to keep it to themselves, for fear their colleagues would laugh at them.

Many developments in the study of the mind have happened since their work together first appeared. Some scientists now assert that it should be possible to simulate the working of the mind on a computer, encapsulated as logical procedures for solving problems. This echoes the logical empiricists of the Vienna Circle in the early twentieth century.

Neurophysiologists such as Antonio Damasio investigate how parts of the brain are stimulated by images or problems. One of the instruments they use to study which sections of the brain become stimulated and in what order is functional magnetic resonance imaging (fMRI). This produces images of the brain derived from the way electrons line up their spins in a magnetic field, the understanding of which derives from Pauli's fourth quantum number. Another method is to measure the increased oxygen flow to a particular area of the brain that occurs when a subject solves a task. To do so the researcher injects radioactive oxygen into the subject's bloodstream. It produces positrons that collide with the electrons in the brain to produce light quanta which are detected by radiation counters around the patient's head (proton emission tomography [PET]). These are examples of the marriage of physics and brain research that Pauli might have thought up himself.

Neurophysiologists have also produced evidence showing how important visual images are to the working of the mind. Pauli's discovery of the fourth quantum number brought to an end the convenient image of the atom as a miniature solar system and led both him and Heisenberg to decide reluctantly to abandon the use of visual images. But Pauli hoped that some day, somehow, in a new theory, a usable visual image of atomic processes would be discovered. Richard Feynman's theory of quantum electrodynamics, formulated in 1949, produced just such an image. Feynman produced the Feynman diagrams, deduced from the equations of his new quantum electrodynamics which was free of the infinities that had rendered invalid Pauli's and Heisenberg's theory of quantum electrodynamics of the 1930s.

Pauli was well aware of Feynman's theory and was not satisfied because it concerned only electrons and light. It did not remove the infinities from theories that contained newly discovered elementary particles, nor those in Fermi's theory of weak interactions; and it did not produce the fine structure constant, either. Pauli did, however, agree that Feynman's procedure—which physicists called "renormalization"—was along the right lines. Jokingly he referred to a Feynman diagram as a "sentimental painting."

In recent years Roger Penrose has made a pioneering attempt to combine neurophysiology with physics. He suggested that structures within neurons—microtubules—could be the seat of the quantum computations that are the dynamics behind thinking. But these are not simply logical computations, because quantum physics, which includes the uncertainty principle and the concept of ambiguity, introduces an indefinable extra ingredient—intuition. This element is not addressed in any of the research programs mentioned here and is a critical shortcoming. Pauli and Jung emphasized its importance, as did Einstein and other scientists when they recalled how they had made their discoveries.

The puzzle of how we reason, how we think—of how we create knowledge from already existing knowledge and how we draw conclusions that go beyond the premises—cannot be solved by logic alone.

Researchers in cognitive science have applied a cross-disciplinary approach. This includes simulating the mind on a digital computer; neurophysiology; notions from philosophy applied to the mind (philosophy of mind); linguistics (how metaphors arise and how they are used); and visual imagery (how visual images are generated and manipulated in problem solving). But they fail to include physics. And despite applying so much heavy intellectual machinery to the study of how the mind operates, they also omit data from the history of science in the form of testimonies, correspondence, and other biographical details of scientists themselves.

In this field Pauli's application of Jung's psychology to Kepler's thinking is an exemplary work. Applying data from case histories of great scientists as grist for the mill of theories of psychology is an adventurous and fruitful route.

In turn, Jung's psychology can throw light on how Pauli made his first great discovery of the exclusion principle: input from his conscious thinking energized the archetypes for three and four (constellated them, to use Jung's terminology), which sparked his insight. Jung, too, found number archetypes essential in transforming the neuroses in his life into a creative force.

Which brings us to 137 and Pauli's obsession with deriving the fine structure constant from quantum electrodynamics. Not only does this remain unsolved but the problem has widened. In Pauli's day there were seven known fundamental constants. Now there are twenty-six. This is due to the increase in the number of known elementary particles, their fundamental interactions, and their properties. While Pauli was able to focus on the fine structure constant and quantum electrodynamics, physicists are now trying to derive all twenty-six from a theory that will encompass not only the electromagnetic force—which is controlled by the fine structure constant—but the strong and weak forces, and eventually the gravitational force as well. This is the ultimate ambition of string theorists, among others whose grand aim is to come up with a theory that explains the large and the small, the universe and the atom—a theory of everything.

Notes

Author's Note

To avoid unnecessary duplication, I use abbreviations of the sort "Pauli (1952), p. 253," which means the work listed in the Bibliography (pp. 311–320) under Pauli, dated 1952, with the quotation cited on p. 253.

Letters from *PLC* are cited according to the code "Pauli to Weisskopf, January 17, 1957: *PLC7* [2445]," which is the letter Pauli wrote to Weisskopf on January 17, 1957. [2445] means it is the 2,445th letter in Pauli's published correspondence and also stands for the letter's catalogue number in *PLC7*, which is the volume containing Pauli's correspondence from 1957.

The correspondence between Pauli and Jung in *P/J* is cited according to the code "*P/J* [76P], August 5, 1957." [76P] designates the 76th letter written by Pauli [P] to Jung, dated August 5, 1957.

Unless indicated otherwise, all use of italics in quotations occurs in the original text.

Prologue

xxi "difficult transition from three to four": Pauli to Fierz, October 3, 1951: *PLC4* [1286].

xxii **"against the rationalism of the eighteenth century"**: Sommerfeld (1927), p. 195.

xxii **"connects with the old mystic elements"**: Pauli to Hertha Pauli, October 11, 1957: *PLC7* [2707].

xxiii **"with his Prism and silent Face"**: William Wordsworth, *The Prelude*, Book Three, line 59.

xxiii **"He was the last magician"**: Keynes (1947), p. 27.

xxiii **"in the doctrines of alchemy"**: Brewster (1831), p. 271.

xxiv **two or three flips out of ten thousand**: See Benedict Carey, "A Princeton Lab on ESP Plans to Close Its Doors," *The New York Times*, February 10, 2007.

xxv **"in principle, the *whole* world"**: Pauli (1952), p. 259.

xxv **"the darkest hunting ground of our times"**: Jung to Ira Progoff, January 30, 1954, copy at the ETH; quoted in Bair (2004), p. 553.

Chapter 1 · Dangerously Famous

3 **"fur-coat ladies"**: See Bair (2004), p. 98 and p. 683, private communication to her.

3 **"like a large genial cricketer"**: Hayman (1999), p. 300.

3 **"powerful arms"**: Elizabeth Shepley Sergeant, "Dr. Jung: A Portrait in 1931," *Harper's*, May 1931.

3 **"dangerously famous"**: Jung to Bailey, November 25, 1932. From Jung and Ruth Bailey's correspondence, ETH; quoted from Bair (2004), p. 401; private communication, Samuel Beckett to Thomas McGevey, undated note, postmarked January 1935.

3 **"going to Jung was somehow very chic and modern"**: Jung to Bailey, November 25, 1932. From Jung and Ruth Bailey's correspondence, ETH; quoted from Bair (2004), p. 401; private communication, Samuel Beckett to Thomas McGeevy, undated note, postmarked January 1935.

4 **"without any direct line of tradition"**: *MDR*, p. 38.

4 **"a tremendous experience for me"**: *MDR*, p. 122.

5 **"where nature would collide with spirit"**: *MDR*, p. 130.

5 **"the So-Called Occult Phenomena"**: *CW1*, pp. 3–88.

6 **The friend, sadly to say, laughed**: See Hayman (1999), p. 36.

8 **"heroic efforts"**: *F/J* [265J], July 19, 1911.

8 **"license to be unfaithful"**: *F/J* [175J], January 30, 1910.

9 **"thin close-cropped hair"**: Quoted from Gay (1989), p. 202.

10 **he tried a "talking cure"**: Anna O., a patient of Breuer's, coined the term "talking cure" in the early 1880s. See Gay (1989), p. 63.

10 **"dabbling in spookery again"**: *F/J* [50J], November 2, 1907.

10 **"becoming red-hot—a glowing vault"**: *MDR*, pp. 178–179.

11 **"catalytic exteriorization phenomenon"**: *MDR*, pp. 178–179. Note: British spelling has been corrected to American spelling.

11 **"Sheer bosh"**: *MDR*, p. 179.

11 **divesting "him of any paternal dignity"**: *F/J* [139F], April 16, 1909; also in *MDR*, p. 397.

11 **"what we are bringing to them"**: Interview with Jung's son Franz by Bair, in Bair (2004), pp. 159 and 700 (note 92).

11 **"personal authority above truth"**: *MDR*, pages, 181–182.

12 **"spiritual aspect and its numinous meaning"**: *MDR*, p. 192.

12 **she had confessed it to him:** Over the years Freudians dismissed Jung's claim as malice. But in December 2006 a German researcher happened to read the ledger of a tiny hotel tucked away in the Swiss Alps. In Freud's distinctive handwriting were the words "Dr Sigm Freud u frau [Dr Sigmund Freud and wife]," but, in fact, the woman with him was not his wife but Minna Bernays, his sister-in-law. Freud sent a postcard to his wife describing the beautiful scenery the two of them had seen. Ralph Blumenthal, "Hotel Log Hints at Illicit Desire that Dr. Freud Didn't Suppress," *The New York Times*, December 24, 2006.

13 **began to enter his dreams:** *MDR*, p.189.

14 **the personal unconscious and the conscious:** In his 1919 lectures in London, "Instinct and the Unconscious," Jung tells us that he "borrowed the idea of archetype" from writings of Saint Augustine. (*CW8*, pp. 135-136). It also appears in Gnostic literature, with which Jung was familiar from his university days.

14 **namely, the collective unconscious:** See *CW8*, p. 372.

14 **"new life to the people to whom they came"**: Foreword to the Swiss edition of *CW12*, p. vii.

15 **"images of my own unconscious"**: *MDR*, p. 233.

15 **"and reads to him from a scroll"**: *MDR*, p. 38.

16 **illustrated in the manner of a medieval manuscript:** *MDR*, p. 213.

16 **"Jung was a walking asylum in himself"**: Hull to Savary, June 20, 1961, quoted from Bair (2004), pp. 292–293.

Chapter 2 · Early Successes, Early Failures

18 **a few minutes at Göttingen station:** Gamow (1985), p. 64.

19 **Wolfgang Ernst Friedrich Pauli:** The name Friedrich came from his maternal grandfather Friedrich Schütz, who often took the young boy on long walks around Vienna.

20 **" 'Antimetaphysical ancestry' "**: *P/J* [60P] March 31, 1953; translation from von Meyenn and Schucking (2001), pp. 43–44.

20 "flows into the Danube Canal": Airmail letter, January 1959, from Hertha to Wolfgang, CERN Archive Collection, Document PLC Bi 120.

20 played "Silent Night" on the piano: Autobiographical statement by Hertha Pauli, in *Guide to Catholic Literature*, CatholicAuthors.com.

21 Pauli said no, but Hertha insisted: Peierls (1960), p. 175; and Cline (1987), pp. 136–137.

21 *das U-boot* (the submarine): Enz (2002), p. 15.

22 found out about his Jewish ancestry: Interview with Ewald by T. S. Kuhn, *AHQP*, p. 15.

22 a chemist named Pascheles—who was in fact himself: I thank Karl von Meyenn for this story, which had been told to him by Pauli's close friend Marcus Fierz.

22 speaking of herself as "half-Christian": Communication from Drs. Susanne Blumesberger and Ursi Gabl, who have been researching Hertha Pauli's papers at the Austrian National Library. See also Hertha Pauli's biographical statement at CatholicAuthors.com. Even her publisher in New York City, Farrar, Straus & Giroux, was under the impression she was Catholic. See New York Public Library Manuscripts and Archives Division, Farrar, Straus & Giroux Inc. Records: Box 273 (General Correspondence Folder), particularly office memos from June 1970.

22 "With me everything is complicated": Pauli to Jaffé, November 28, 1950: *PLC4* [1172].

23 little personal correspondence: See the editorial note by Karl von Meyenn in *PLC8*, p. 1376.

23 the most beautiful theory ever formulated: In 1905 Einstein had discovered the special theory of relativity. It is "special" or restricted in that it covers only measurements made in laboratories moving at a constant speed in a straight line and does not include gravity. General relativity includes measurements made in laboratories moving in every direction, including a circle.

24 a paper on relativity theory: Pauli (1919).

24 increased as the war went on: *PLC1*, p. xxiv.

24 "There I have indeed made a mistake": Jordan (1971), p. 33.

26 "I have with me a really astonishing specimen": Sommerfeld to von Geitler, January 14, 1919; for facsimile go to www.irz-muenchen.de/~Sommerfeld/gif100/0584_01.gif.

26 "I find it impossible to understand how": Weyl to Pauli, May 10, 1919: *PLC1* [1].

27 "the sureness of critical appraisal": Einstein (1922), p. 184.

27 at "the moment meaningless for physics": Pauli to Eddington, September 20, 1923: *PLC1* [45].

27 "not really as stupid as it may have sounded": Hoyle (1994), p. 310. See also *PLC8*, pp. 1205–1206.

27 **"laboratory fellow sufferer"**: Pauli to Sommerfeld, September 29, 1924: *PLC1* [64].

28 **"eight o'clock in the morning"**: Interview with Heisenberg by T. S. Kuhn, *AHQP*, November 11, 1963, p. 28.

31 **"a simple farmboy"**: Born (1975), p. 212.

31 **"it was nothing serious at all"**: Interview with Heisenberg by T. S. Kuhn, *AHQP*, November 30, 1962, p. 3.

32 **"with great concentration and success"**: Heisenberg (1971), p. 27.

32 **"That helped me a lot"**: Interview with Heisenberg by T. S. Kuhn, *AHQP*, November 30, 1962, p. 8.

32 **"I believed I was the best formalist of my time"**: Interview with Pauli by Jagdish Mehra, February 1958, in Mehra and Rechenberg (1982), p. xxiv.

36 **"toward the knowledge of the unity of the laws of the world"**: Born (1923), p. 537.

37 **"this confidence [in Bohr's theory] was shaken"**: Interview with Heisenberg by T. S. Kuhn, *AHQP*, November 2, 1963, p. 8.

37 **"W. Pauli is now my assistant"**: Born to Einstein, October 21, 1921, in Born (1971), p. 58.

37 **"cannot bear life in a small city"**: Born to Sommerfeld, January 5, 1923, in Sommerfeld (2004), p. 137.

38 **"until the small hours of the morning"**: Born (1975), p. 212.

38 **"I shall never get another assistant as good"**: Born to Einstein, November 29, 1921, in Born (1971), pp. 61–62.

38 **"He also plays the piano very well"**: Born to Einstein, April 7, 1923, in Born (1971), p. 75.

38 **"All existing He[lium]-models are false"**: Heisenberg to Pauli, March 26, 1923: *PLC1* [34].

38 **"I met Niels Bohr personally for the first time"**: Pauli (1946), p. 213.

39 **"Göttingen's famous school of mathematics"**: Heisenberg (1971), p. 38.

39 **"somewhat Kabbalistic"**: Sommerfeld (1923), p. 59; see also Pauli (1946), p. 166.

40 **no one went to bed before 1 a.m.**: Heisenberg to Karl and Helen Heisenberg, June 15, 1922, quoted from Cassidy (1992), p. 128.

40 **"far exceeds my abilities"**: *CSP*, Volume II, p. 1073.

40 **"It was given by Nature herself without our agency"**: Sommerfeld (1923), p. 237. In atoms with one electron in their outer shell—such as hydrogen and sodium—each of their spectral lines is split into two, called doublets. In atoms with two electrons in their outer shells, each line is split into three—triplets. Scientists only discovered the reason for this when spin was postulated.

41 **"a revelation"**: Einstein to Sommerfeld, February 8, 1916, in Hermann (1968), p. 40.

41 **"than your beautiful work"**: Bohr to Sommerfeld, March 1916, in Hoyer

(1981),Volume 2, p. 603. See Kragh (2003) for further historical details concerning the fine structure constant.

42 or would be dramatically different: Barrow (2003), pp. 154–156.

43 "thinking about the anomalous Zeeman effect": *CSP*,Volume II, p. 1073.

Chapter 3 · The Philosopher's Stone

46 retrieve and develop these inferior functions: See Jung (1921), pp. 563–565; and *CW18*, pp. 16–18.

47 "off the beaten track and rather silly": *MDR*, p. 230.

47 "those old acquaintances of mine": *MDR*, pp 230–231.

47 then return to Silberer's and Wilhelm's books: Freud's staunch refusal even to entertain Silberer's ideas on alchemy drove Silberer to suicide in 1923.

49 "psychology of the unconscious": *MDR*, p. 231.

49 when alchemy was at its height: *MDR*, p. 231.

50 "Don't waste your time": Interview with Aniela Jaffé by Ean Beg, July 27, 1975, presented on BBC; quoted from Bair (2004), p. 379. Jung appointed Aniela Jaffé secretary of the institute. In 1931 she had started a degree in psychology at the University of Hamburg but was forced to flee and went to Zürich. After undergoing analysis with Jung, she became intensely interested in his analytic psychology and wrote many papers on it. In 1955 she became his personal secretary and collaborator. She screened his incoming correspondence as well as informed him of interesting ideas in letters addressed to her. She was an indirect path to Jung, of which Pauli often availed himself.

Chapter 4 · Dr. Jekyll and Mr. Hyde

51 "I feel myself so unwell": Pauli to Bohr, December 12, 1923: *PLC1* [43].

52 Surely it was all tied together: Pauli (1946), p. 168.

52 "We will see what you can do": Interview with Heisenberg by T. S. Kuhn, *AHQP*, November 30, 1962, p. 5.

53 "Success sanctifies the means": Heisenberg to Pauli, November 19, 1921: *PLC1* [16].

54 "with a tear in my eye": Pauli to Sommerfeld, June 6, 1923: *PLC1* [37].

54 "unsightly" and "monstrous": The quotations are from Pauli to Alfred Landé, December 14, 1923: *PLC1* [51]; and Pauli to Kramers, December 19, 1923: *PLC1* [52], p. 135.

54 "I am deeply insulted by it": Pauli to Bohr, February 11, 1924: *PLC1* [54].

54 "do it again with halves": Pauli to Bohr, February 21, 1924: *PLC1* [56].

54 "we have to create something fundamentally new": Pauli to Landé, August 17, 1923: *PLC1* [42].

54 "no taste at all for this sort of theoretical physics": Pauli to Bohr, February 21, 1924: *PLC1* [54].

54 "too difficult": Pauli to Bohr, February 11, 1924: *PLC1* [54].

55 "particularly if they are women": Pauli to Wentzel, December 5, 1926: *PLC1* [149].

55 "without love, indeed without humanity": Pauli to Rosebaud, December 13, 1955: *PLC6* [2214].

55 "in my relations with women": *P/J* [69P], October 23, 1956.

55 Hamburg welcomed her with open arms: Sauvage (1949), pp. 124, 131–132.

56 "outbursts of ecstasy and visions": *P/J* [30P], May 24, 1934.

56 hoping she was gone forever: This is based on *P/J* [69P], October 23, 1956; Pauli to Rosenbaud, December 13, 1955: *PLC6* [2214]; and Pauli to Fierz, March 2, 1956: *PCL6* [2253].

57 "Moulin-Rouge, or something analogous": Pauli to Wentzel, May 16, 1927: *PLC1* [162].

57 chalked it up to the Pauli effect: Schucking (2001), pp. 46–47.

57 "was not allowed to enter": Interview with Stern by Res Jost, November 11 and December 2, 1961, p. 38; on deposit at the ETH Bibliothek.

59 a physicist at Cambridge University: Dirac (1926), p. 670.

60 understanding the structure of the atom: Pauli to Bohr, December 12, 1924: *PLC1* [74].

61 "swindle x swindle does not yield something correct": Heisenberg to Pauli, December 15, 1924: *PLC1* [76].

61 "complete insanity": Bohr to Heisenberg, December 22, 1924: *PLC1* [77].

61 "I would be the happiest man on earth": Pauli to Bohr, December 31, 1924: *PLC1* [79]. For detailed studies of Pauli's discovery of the exclusion principle, see Heilbron (1983), Hendry (1984), Jammer (1966), Massimi (2005), Mehra (1982), and von Meyenn (1980, 1981).

62 "then will visualizability be regained": Pauli to Bohr, December 12, 1924: *PLC1* [74].

62 "indeed a witty idea": Kronig (1960), p. 21.

Chapter 5 · Intermezzo—Three versus Four

64 "in spite of manifold variety": Preface to the 1919 edition of Sommerfeld (1923), p. viii.

65 "somewhat Kabbalistic": Sommerfeld (1923), p. 59.

65 "or the witches' kitchen of Faust": Andrade (1926), p. 708.

65 **"For integers, go to Sommerfeld"**: Interview with Heisenberg by T. S. Kuhn, *AHQP*, October 30, 1962, p. 12; and Pauli (1948b), p. 65.

65 **"Kepler's magnum opus—*Harmonices Mundi*"**: Pauli (1948b), p. 68.

65 **he may well have read him in the original Latin:** Pauli's Latin was good enough to at least get the gist of Kepler's text. He had studied Latin for eight years at the gymnasium with grades ranging over "very good," "good," and "sufficient." I thank Karl von Meyenn for a copy of Pauli's grades at the Döblinger Gymnasium.

65 **"I, myself, am not only Kepler, but also Fludd"**: Pauli to Fierz, January 19, 1953: *PLC5* [1507].

66 **"That was really the *main work*"**: Pauli to Fierz, October 3, 1951: *PLC4* [1286].

66 **"philosophical problem connected with these numbers"**: Pauli to Fierz, October 3, 1951: *PLC4* [1286].

66 **"origin of and development of concepts and theories"**: Pauli (1952), p. 220.

67 **"left a large sore"**: Quoted from Koestler (1964), p. 233.

67 **"pains of the bladder"**: Quoted from Koestler (1964), p. 234.

69 **"the planets which circle around him"**: Quoted from Kuhn (1957), p. 131.

69 **"such as one cannot find elsewhere"**: Quoted from Caspar (1993), p. 91.

70 **"Three, yet are they One"**: Quoted from Pauli (1952), p. 230.

70 **"from the very beginnings of mankind"**: Quoted from Pauli (1952), p. 227.

70 **"Geometry is the archetype of the beauty of the world"**: Quoted from Pauli (1952), p. 223.

71 **"imitate the sun"**: Quoted from Pauli (1952), p. 231.

71 **"the true laws of planetary motion"**: Quoted from Pauli (1952), p. 232.

71 **"for the creation of the universe"**: Quoted from Pauli (1952), p. 228.

72 **"source and root of eternal Nature"**: Quoted from Pauli (1952), p. 141.

75 **"divine ordinance"**: Quoted from Caspar (1993), p. 62.

75 **"Copernicus had told the truth"**: Quoted from Caspar (1993), p. 62.

77 **straighten it out in eight days:** Caspar (1993), pp. 126–127.

77 **the nova of 1604:** What Kepler saw was, in fact, a supernova.

78 **also fitted Mars's measured eccentricity:** Caspar (1993), p. 139.

78 **and sea water its nourishment:** Pauli (1952), p. 235.

79 **which he believed to exist in nature:** Kepler to von Wackenfels, 1618, quoted from Caspar (1993), p. 265.

80 **"perceived by the intellect, not by the ear"**: From Kepler's *Harmonices*, quoted from Koestler (1964), p. 398.

80 **"*Misery and Famine reign on our planet*"**: Quoted from Caspar (1993), p. 284.

80 **"for one to contemplate His works"**: Quoted from Caspar (1993), p. 343.

82 **"by means of pictures"**: Quoted in Westman (1984), p. 179.

82 **"comprehend the true core of natural bodies"**: Quoted from Pauli (1952), p. 251.

82 **"mine is the task of the mathematician"**: Quoted from Koestler (1964), p. 403.

83 **"murky mirror of the world drawn underneath"**: Quoted from Pauli (1952), p. 244.

83 **"triangle seen in the mirror of the world"**: Quoted from Pauli (1952), p. 244.

85 **"kind of mystic primordial space"**: Scholem (1941), p. 261. See also Dan (2006).

85 **"This is alchemy in the best sense"**: Pauli to Fierz, January 19, 1953: *PLC5* [1507], p. 20.

85 **"a world drawn in pictures"**: Quoted from Westman (1984), p. 206.

85 **"I am like a blind man"**: Quoted from Pauli (1952), p. 252.

86 **adds up to the magic number seven**: Quoted from Pauli (1952), p. 270.

86 **transformations that produce our world**: Quoted from Pauli (1952), pp. 274–275.

86 **"it is astonishing how much smoke they expel"**: Kepler to Seussius, February 28, 1624; quoted from Caspar (1993), p. 293. See Westman (1984) for more on the Kepler-Fludd polemic. See also Holton (1973).

87 **King James I to Germany**: For details, see Bernstein (1997).

87 **"the eternal fountainhead of nature"**: Quoted from Pauli (1952), p. 271.

88 **"I have certain features of both"**: Pauli to Fierz, October 3, 1951: *PLC4* [1286].

Chapter 6 · Pauli, Heisenberg, and the Great Quantum Breakthrough

90 **"Bohr will rescue us with a new idea"**: Pauli to Kronig, May 21, 1925: *PLC1* [89].

90 **"he will greatly advance science"**: Pauli to Bohr, February 11, 1924: *PLC1* [54].

91 **"We must adjust our concepts to experience"**: Pauli to Bohr, December 12, 1924: *PLC1* [74].

91 **"a renewed enjoyment in life"**: Pauli to Kronig, October 5, 1925: *PLC1* [100].

93 **"it is not the only [philosophical approach]"**: Pauli to Schlick, August 21, 1922: *PLC2* [39A].

93 **"There is no logical path to these laws"**: Einstein (1918), p. 4.

94 **"from the physicist's mode of thinking"**: Mach (1910), p. 37.

94 **"to any God, authority or 'ism' "**: Pauli to von Franz, February 17, 1955: *PLC6* [2019].

94 **"will *visual imagery* be regained"**: Pauli to Bohr, December 12, 1924: *PLC1* [74].

95 **Bohr applauded Pauli's "wonderful results"**: Bohr to Pauli, November 25, 1925: *PLC1* [109].

95 Pauli had done it "so quickly": Heisenberg (1960), p. 40; and Heisenberg to Pauli, November 3, 1925: *PLC1* [103].

95 they had the theory right: Heisenberg and Jordan (1926).

96 "and by the lack of visualizability": Schrödinger (1926), p. 735.

97 "visualizability of his theory I consider crap": Heisenberg to Pauli, June 8, 1926: *PLC1* [136].

97 how the two sets of spectra arise: Heisenberg (1926). See Miller (1995), pp. 9–12 for details.

99 "who is calculating H_2^+ according to Schrödinger": Pauli to Wentzel, June 11, 1926: *PLC1* [138].

99 wave functions that Burrau had deduced: Heisenberg to Pauli, November 23, 1926 [148]. See Burrau (1926/1927).

99 "state of almost complete despair": Heisenberg to Pauli, November 23, 1926: *PLC1* [148].

100 in the end Born took the credit: Pauli included it as footnote to one of his papers on magnetism—Pauli (1927), p. 83.

100 "every time I reflect on it": Heisenberg to Pauli, November 4, 1926: *PLC1* [145]; the letter Heisenberg referred to is Pauli to Heisenberg, October 19, 1926: *PLC1* [143].

101 apply such words with great care: To give you a taste of the weirdness of quantum mechanics, the "numbers" it uses are of a nonstandard sort. So non-standard that when Heisenberg published his original paper on the quantum mechanics he, himself, was confused. He found that when he multiplied the x- and y-coordinates for the position of a particle as xy, it was not the same as the value for the reverse order, yx—mathematicians say that in this case the property of commutativity does not apply: xy is not the same as yx. The numbers we deal with in our daily life possess the property of commutativity, which means that $3 \times 2 = 2 \times 3$. Numbers like 2 and 3 commute.

But this is generally not so for quantum mechanics, where the mathematical symbols for position (Q) and momentum (P) do not commute. It boils down to the appearance of Planck's constant. If Planck's constant were zero, then Q and P would commute, that is, $QP = PQ$. Rather in quantum mechanics the relevant equation is: $PQ - QP = \dfrac{h}{2\pi i}$.

101 who had given him the key idea: Heisenberg to Pauli, February 23, 1927: *PLC1* [154].

101 "It becomes day in the quantum theory": Heisenberg (1960), p. 40.

101 light and electrons behaved like particles: Heisenberg had concluded that collisions between electrons and light quanta were the root of uncertainties in any measurement of the electron's position and momentum. In this way he missed the critical point of examining how the accuracy of the measurement of the position of an electron is limited by how a microscope resolves the light entering its eyepiece.

To give more depth to the uncertainty principle, Bohr improved on Heisenberg's method of deducing the uncertainty relations by analyzing how the *wavelength* of the light bouncing off an electron is measured by a microscope.

In this way Bohr showed how important the wave-particle duality of light and electrons was in deducing the uncertainty principle.

102 **context of waves and particles:** Bohr went on to elucidate the critical role that Planck's constant played in measurements because if Planck's constant were zero, then there would be neither a wave-particle duality nor an uncertainty relation.

An electron's momentum (an aspect of its particle nature, p), and its wavelength (an aspect of its wave nature, λ) are related by Planck's constant h as follows: $p = h/\lambda$. According to Heisenberg's uncertainty relation, the product of the error in the measurement of its position (Δx) and the error in the measurement of its momentum (Δp) is $\Delta x \Delta p > h/2\pi$. Although Planck's constant h is very small—a tenth of a billionth of a trillionth of a trillionth—it is not zero. If it were zero then the electron's momentum and wavelength would be unrelated (there would be no more wave-particle duality) and the uncertainty relation would disappear as well.

This can be summarized schematically:

$$\text{wave} \leftarrow (h) \rightarrow \text{particle}$$
$$\text{error in position measurement} \leftarrow (h) \rightarrow \text{error in momentum measurement}$$

In other words, Planck's constant determines the relationship between wave and particle and between the error in measurement of the position and momentum of an electron. So when if h (Planck's constant) is zero, we return to the world of our daily experiences in which there is wave-particle duality and there is no uncertainty relation.

This led Bohr to conclude that when any measurement is carried out in the world of atoms, the system undergoing measurement (in this case the electron) and the measurement system (the light which strikes it and caroms into a microscope) are inextricably linked, changing the properties of the system being measured in ways that cannot be exactly determined. In Newtonian science, on the other hand, we do not have to take into consideration the effects of light hitting a falling stone as we observe it.

102 **"distinction between subject and object":** Bohr (1961), p. 91.

102 **he entirely agreed with Bohr's thesis:** Bohr to Pauli, October 11, 1927: *PLC1* [172]; Pauli to Bohr, October 17, 1927: *PLC* [173].

103 **"wave and light quantum descriptions":** Dirac (1927), p. 245.

103 **"saddest chapter in modern physics":** Heisenberg to Pauli, July 31, 1928: *PLC1* [204].

104 **" 'I should have taken Bethe' ":** Interview with Weisskopf by Karl von Meyenn, July 10, 1963, in *PLC2*, p. xxi.

105 **"applications of special relativity theory":** Pauli (1940), p. 722.

Chapter 7 · Mephistopheles

107 "And nobody understood anything": Interview with George Uhlenbeck by T. S. Kuhn, *AHQP*, March 30, 1962, p. 5.

107 "I like your publications better than I like you": Cline (1987), p. 138.

107 "only ONE God's whip (Thank God!!!)": Ehrenfest to Pauli, November 26, 1928: *PLC1* [211].

108 at the appropriate time, use it: Interview with T. D. Lee by the author, Columbia University, April 23, 2008.

108 unified theory of gravitation and electromagnetism: Pauli to Einstein, December 19, 1929: *PLC1* [239].

108 "So you were right, you rascal": Einstein to Pauli, January 22, 1932: *PLC2* [288].

108 "not even accorded Bohr": Pauli to Sommerfeld, December 2, 1938: *PLC2* [537a].

108 "may I formulate it this way": Weisskopf (1989), p. 160. *Herr Geheimrat* is usually translated as "Privy Chancellor." In this case Pauli meant it as "Honored Teacher."

108 "not sung to me in the cradle": Pauli to Pais, August 17, 1950: *PLC4* [1147].

108 "without having Pauli read it first": The first quote is from Heisenberg to Pauli, November 21, 1925: *PLC1* [107]; the second is from Hermann (1979), p. XLII.

108 "conscience of physics": Weisskopf (1989), p. 159.

109 "wicked stepmother": *P/J* [69P], October 23, 1956.

110 "which one had to find (and polish) oneself": Telegdi (1987), p. 433.

110 "demonic aura surrounding this queer man": Quoted from von Meyenn (2007), p. 248.

110 "with a rucksack on my back": Pauli to Weyl, November 9, 1955: *PLC1*, p. 443.

111 "Scherrer circus": Enz (2002), p. 198; and interview with Igal Talmi by the author, Weizmann Institute, Rehovot, Israel, January 24, 2007.

111 "it is all right, but also wrong": Weisskopf (1989), p. 160.

111 "contradict me with detailed arguments": Pauli to Kronig, November 22, 1927: *PLC1* [175].

112 "reinvigorate my interest in physics": Pauli to Bohr, January 16, 1929: *PLC* [214].

113 "I would like to meet you. Scherrer": Postcard written June 4, 1928: *PLC1* [199]. In 1933 Jordan openly declared his membership in the National Socialist (Nazi) party and became a Storm Trooper. This was no doubt why he never received a Nobel Prize. During the war Jordan worked for the Nazi advanced-weapons program at Peenemünde.

Afterward Pauli stood up for his former pal "PQ – QP," and urged his rein-
statement into academia, which occurred in 1953, paving the way for Jordan's
return to the University of Hamburg. Soon afterward he was elected to the German parliament in the Adenauer days, as a member of the right-wing Christian
Democratic Party. In this capacity he lobbied for arming the Bundeswehr with tac-
tical nuclear weapons. Born, Heisenberg, and Pauli, among other scientists, strongly
protested. As Pauli pungently put it, "Alas, good Jordan! He has served all regimes
in utmost faithfulness" (Enz [2005], p. 47).

113 **"he is already asserting the opposite"**: Peierls (1985), pp. 46–47.

113 **"and psychoanalysis as a vocation"**: Quoted from Pais (2006), pp. 17–18.

114 **"Not for the curious"**: Wentzel (1960), p. 51. The two papers are Heisenberg
(1929, 1930). See Miller (1995), chapter 3, for details.

114 **"rather introspective—i.e., Buddha-like"**: Pauli to Kronig, December 22,
1949: *PLC3* [1067].

114 **"Jew from the waist up"**: Interview with Igal Talmi by the author, Weizmann
Institute, Rehovot, Israel, January 24, 2007.

116 **married only in a very "loose way"**: Pauli to Klein, March 10, 1930: *PLC2*
[243a].

116 **"This gave him great satisfaction"**: Interview with Käthe Deppner by Jagdish
Mehra, March 12, 1974, in Mehra and Rechenberg (1982), p. xxxvii.

117 **"but with an average chemist"**: Interview with Franca, who was Pauli's second
wife, by Charles Enz, March 21, 1971, in Enz (2002), p. 211.

117 **"vehemently plagued by jealousy"**: Weyl to Hecke, May 28, 1930: *PLC5*, p. xxi.

117 **Käthe married her chemist, Paul Goldfinger:** After the war, from time to
time, Käthe used to show up unexpectedly at Pauli's home in the small town of
Zollikon, just outside Zürich. Franca did not at all appreciate this and refused to let
her in, so Pauli went on long walks with her. Interview with Karl von Meyenn by
the author, November 14, 2006.

117 **plus the discharged electron:** At the time physicists were debating about
whether there were electrons as well as protons inside the nuclei of atoms. But this
model of the nucleus led to certain properties that clashed with those measured in
the laboratory. The discovery of the neutron, in 1932, clarified the situation. The
neutron has about the same mass as the proton and the same spin, but it has no
electric charge. In beta-decay the neutron transforms itself into an electron, a proton, and a neutrino, and the electron is immediately expelled from the nucleus.

117 **"We must still be prepared for new surprises"**: Bohr (1930), p. 371.

117 **"that something was missing"**: Interview with Igal Talmi by the author, Weizmann Institute, Rehovot, Israel, January 24, 2007.

118 **"which further behaved foolishly"**: Pauli to Delbrück, October 6, 1958: *PLC8*
[3075].

118 **a ball at the splended Baur au Lac:** Pauli to Meitner et al., December 4, 1930:
PLC2 [259].

119 **"But under 'dryness' I don't suffer at all":** Pauli to Peierls, July 1, 1931: *PLC2* [279].

119 **"until my bones are whole again—very tedious":** Pauli to Peierls, July 1, 1931: *PLC2* [280].

119 **"like a traffic cop signalling":** Tippys to Goudsmit, August 19, 1931: *PCL2*, p. 84.

119 **inverse Pauli effect:** Quoted from *PCL2*, p. 84.

120 **raised his hand in a *"Heil Hitler"* salute:** Quoted from *PLC2*, p. 84.

120 **"second-order acquaintances":** Pauli to Wentzel, September 7, 1931: *PLC3* [283a].

120 **"very simple and very neat here":** Pauli to Wentzel, September 7, 1931: *PLC3* [283a].

120 **"Your old Pauli":** Pauli to Wentzel, September 7, 1931: *PLC3* [283a].

121 **"two problematic temperaments":** Pauli to Weisskopf, December 30, 1940: *PLC3* [615].

121 **"increase her publicity with your help":** Pauli to Delbrück, October 6, 1958: *PLC8* [3075].

121 **infuriating grin and a long tail:** Gamow (1985), pp. 165–214.

123 **delighted to be cast in this role:** Schucking (2001), p. 46. For more on this spoof, see Segré (2008).

Chapter 8 · The Dark Hunting Ground of the Mind

124 **three for extremely important:** Pauli's personal library is housed in *La Salle Pauli* in CERN.

125 **Jung would help him discover it:** Jung (1921), p. 594.

125 **"a vague dread of the other sex":** Jung (1921), pp. 487–489.

126 **"has caught me and I can deal with it":** Quoted from Joseph Henderson's conversations with Jung, "Carl Gustav Jung: 1875–1961," compiled and presented by Ean Begg, BBC Radio 3, July 27, 1975.

127 **"he made up his mind to consult me":** Jung (1936b), p. 6.

127 **"blowing over from the lunatic asylum":** Jung (1958a), pp. 538 and 540.

128 **visions were driving him to distraction:** See Jung (1944), p. 42; and Jung (1935), pp. 173 and 540.

128 **"Therefore I won't touch it":** Jung (1935), p. 174.

128 **"appropriate man to treat me medically":** Pauli to Rosenbaum, February 3, 1932, ETH Wissenschaftliche Sammlungen, ETH-Library, Hs. 176.

128 **"and so he had spontaneous fantasies":** Jung (1936b), pp. 7 and 8.

129 **"He cannot be inferior":** Jung (1959), p. 14.

129 **"for having to read all this":** Pauli to Rosenbaum, August 26, 1932, ETH Wissenschaftliche Sammlungen, ETH-Library, Hs. 176. There are no letters from

Rosenbaum to Pauli in the archives. Perhaps Franca destroyed them after Pauli's death as she did with those of other women with whom Pauli corresponded.

130 **"marvelous series of archetypal images":** Jung (1935), p. 174.

130 **others suspected it too:** Personal communication from Fierz to Lindorff, see Lindorff (2004), p. 52.

130 **referred to were indeed Pauli's:** See Westman (1984), pp. 217–218. While analyzing Pauli's essay on Kepler, Robert Westman, a historian of science, became curious as to whether Pauli had any close connection with Jung. He asked a colleague in Zürich to contact Meier.

   The information actually could have been found over a decade earlier in a footnote in volume 18, published in 1977, of Jung's *Collected Works*. It was added by the translator, R. F. C. Hull, to a lecture that Jung gave in 1939 in London. (See *CW18*, p. 265 for full citation; the footnote is on p. 285 of Jung's lecture, *The Symbolic Life*.) In his lecture Jung spoke about the case of a "a great scientist, a very famous man, who lives today." Pauli was known to be a colleague of Jung's, and some friends knew they had a rather close relationship. Jung craved the attention of scientists, particularly physicists, and so came close to revealing the forbidden information.

   Jung went into more detail in seminars in the United States during 1936 and 1937—see Jung (1936b, 1937a, 1937b). In 1944 Jung published a lengthy version of his dream analysis in Jung (1944).

131 **"because there is nobody home":** Jung (1937a), p. 96.

132 **"*get away from Father*":** Jung (1944), p. 49.

136 **"and is always the victim":** Jung (1936b), p. 81.

136 **"He held my both hands and kissed me":** Jung (1936b), p. 81.

137 **he was no longer the center of attention:** *P/J* [16P], February 28, 1936.

137 **share her with another man:** Jung (1936b), p. 79.

139 **"of human knowledge and understanding":** Jung (1958a), p. 540.

139 **"the documentary evidence of his sanity":** Jung (1958a), p. 540.

139 **trance states of shamans and medicine men:** Jung (1916), p. 68.

140 **trapped in a world of phantasmagoria:** Jung (1935), p. 174; and von Franz (1972), pp. 108–111.

140 **"the unconscious as personified by the anima":** Jung (1944), p. 112.

144 **"and you will have the Philosopher's Stone":** Jung (1937b), p. 54.

146 **"completely throttling" the left:** Jung (1944), pp. 154–163.

146 **"*I am at one with myself*":** Jung (1944), p. 172.

146 **"but still not good enough":** Jung (1944), p. 174.

146 **"a certain man of unpleasant aspect":** Jung (1944), p. 177.

147 **"the darkest hunting ground of our times":** Jung to Progoff, January 30, 1954, copy at the ETH; quoted in Bair (2004), p. 553.

Chapter 9 · Mandalas

151 **dreams accompanied by his drawings:** Jung (1944), p. 167.

151 **"period of spiritual and human confusion":** Pauli (1955b), p. 30.

151 **"one revolution of the golden ring":** Jung (1944), pp. 203–204; and Jung (1937b), p. 66.

152 **in the Kabbalah, signifying wisdom:** Jung (1944), p. 206.

153 **made up of forty-nine rotating spheres:** See *CW11*, pp. 68–72.

155 **"produced the impression of 'most sublime harmony' ":** Jung (1937b), p. 72.

155 **"discussed in medieval Christian philosophy":** Jung (1937b), p. 74.

155 **"the existence of an archetypal God-image":** Jung (1937b), p. 59.

155 **"became a perfectly normal and reasonable man":** Jung (1935), p. 175.

155 **"development of symbols of the self":** Jung (1944), p. 215.

156 **"outbursts of ecstasy and visions":** *P/J* [30P], May 24, 1934.

156 **"unless something untoward should arise":** *P/J* [7P], October 27, 1934.

157 **"is something I have since rather lost":** Pauli to Hecke, October 20, 1938: *PLC* [534].

157 **"mix my critical remarks with so much sugar":** Pauli to Born, November 20, 1942: *PLC3* [668].

157 **"and a third not essentially influenced":** *MDR*, p. 165.

Chapter 10 · The Superior Man Sets His Life in Order

158 **several high-level secretarial positions:** Friedrich Adler was a colorful figure. As a young man he had been a promising physicist and a close friend of Einstein's, sharing as well an interest in Socialist politics. As a protest against the Austrian government's institution of an autocratic military regime and its decision to dissolve the parliament in 1916, he walked up to the premier while he was eating lunch and shot him three times in the head. Adler was sentenced to death, despite support from leading figures including Einstein. The sentence was never carried out and he was released after the war.

159 **"Now we marry":** Quoted from Enz (2002), p. 286; from notes of Enz's conversations with Franca Bertram, March 21, April 6, and May 6, 1971.

159 **"I am going to get married also":** Weisskopf (1989), pp. 160–161.

159 **dark archetypes into his consciousness:** *P/J* [29P], April 28, 1934.

159 **Jung was "perfectly correct":** *P/J* [29P], April 28, 1934.

159 **"it secretly made a great impression on her":** *P/J* [30P], May 24, 1934.

160 **"the binding would be good":** Enz (2002), pp. 247–248, notes from Enz's interviews with Franca in 1971.

160 Franca had done him a favor: von Meyenn (1999), p. xxiii.

160 screamed that he wanted to "thrash someone": Enz (2002), p. 287, from conversations with Franca and Adolf Guggenbühl, Jr.

160 "But I never did": Weisskopf (1989), p. 161.

160 "huge piece of work": von Meyenn (1999), p. xxv.

161 "of some interest to the psychologist": P/J [9P], June 22, 1935.

161 "our dream psychology": P/J [19J], March 6, 1937.

161 "radioactive nucleus": P/J [13P], October 2, 1935.

162 Pauli never once mentioned the topic: Weisskopf (1989), p. 165.

162 "by the conventional concept of time": P/J [22P], May 24, 1937; and P/J [23P], October 15, 1938.

163 "3 layers to a four-part object (clock)": P/J [23P], October 15, 1938.

163 about which Pauli had a severe phobia: P/J [29P], April 28, 1934.

163 "of these symbols than I do at the moment": P/J [29P], April 28, 1934.

163 "the 'blond beast' is stirring in its sleep": Jung (1935), pp. 163 and 164.

163 "Wotan the wanderer is on the move": Jung (1936c), p. 180.

164 "a higher potential than the Jewish": Jung (1934), p. 166.

164 "to Germanic and Slavic Christendom": Jung (1934), p. 166.

164 "Freud's brethren"—the Jews: Léon to Greene—members of the tercentenary committee—August 26, 1936. Quoted from Bair (2004), p. 419.

165 "the nightmare on the way to being dreamt": Quoted from Bair (2004), p. 419.

165 "have brought relief to many in distress": From the Harvard tercentenary book as quoted from Bair (2004), p. 421.

165 "lives in indissoluble union with the body": Jung (1936a). The quote is on p. 114.

165 Melville's novel *Moby-Dick*: Aaron (2001), p. 49.

165 Cobb shined them himself: Quoted from Bair (2004), p. 420. This story was related to Bair by an acquaintance of the Cobb family, who relished telling it.

165 "long overdue" book on alchemy: Jung to Jacobi, October 27, 1936: CLI.

166 fearing for his life, left immediately: As told by the American author Philip Wylie, a friend and one-time patient of Jung's. Wylie, however, left no written substantiation of Jung's story in his papers.

167 "really menaced and treated as a Jew": Pauli to Aydelotte, May 29, 1940, in PLC3, p. xxviii.

167 "his fitness for naturalization": Rothmund to Rohn, July 16, 1940, in Enz (1997), document II.31. Rothmund is rumored to be the person who came up with the idea of the "J" stamp as a way to classify Jews crossing the Swiss frontier from Germany. The Nazis went on to use it as a way to identify who was Jewish in Germany and Austria.

167 "Pauli's difficulty was due to a colleague": Enz (2002), p. 338.

168 "best wishes to you in this difficult time": *P/J* [31P], June 3, 1940.

169 passed through the town of Lourdes: Hertha tells this story in her autobiographical account of those years in Pauli (1970).

170 initially been planned for only one year: The funding for his visit, from the Rockefeller Foundation, was scheduled to end in 1942. After some uncertainty, an arrangement was reached whereby Pauli's salary for an extended stay was split between the Institute and the Rockefeller Foundation. See Enz (2002), p. 355.

170 "suffered very much—as for all émigré physicists": Scherrer to Rohn, October 15, 1941, in Enz (1997), document II.48.

170 the department's most important physicist: Personal communications from Professors Karl von Meyenn and Ulrich Mueller-Herold. Later in the war, when it was clear that Germany was losing, Scherrer collaborated with the Office of Strategic Services—the forerunner of the CIA—on a plot to kidnap Heisenberg. See Powers (2000). Scherrer retired from the ETH in 1960. He left no reminiscences and destroyed most of his personal papers.

170 take legal action against the ETH: Pauli to Wentzel, December 30, 1941: *PLC3* [646]; see also the telegraph Pauli sent to Rohn on June 7, 1942 in Enz (1997), document II.62.

170 "The past years have been rather lonesome": Pauli to Casimir, October 11, 1945: *PLC3* [780].

171 "legal complications cannot work on military problems": Oppenheimer to Pauli, May 20, 1943: *PLC3*[671].

171 nothing came of it: See *PLC3*, p. 166.

171 Franca had misgivings about her: Conversations of Karl von Meyenn with Franca Pauli.

173 "possibility of suicide in a desperate moment": From the Dulles Archives, Princeton University. Quoted from Bair (2004), p. 492.

173 "as a footnote to a case of Jung's": Quoted from Bair (2004), p. 495 from a "Report" by Bancroft to Dulles. Dulles's record in the war and its aftermath was exemplary. But instead of as a footnote to the case of Jung, he went down in history as a footnote to the failed invasion of Cuba in 1961. President John F. Kennedy forced him to resign as CIA director, thus ending a career in intelligence that began with a coterie of agents that included Carl Jung.

174 creative potential of the greatest complexity: Jung (1918), p. 14.

174 robbed science and culture of their spiritual foundations: See Drab (2005), p. 54.

174 "I have fallen afoul of contemporary history": Jung to von Speyer, April 13, 1934: *CL*, Volume I.

174 "Well, I slipped up": Jaffé (1971), p. 98.

175 Nobel Prize for his discovery of the exclusion principle: Bohr's name is strikingly absent from the list of people over the years who nominated Pauli for a Nobel prize. Throughout their careers, the relationship between the two men was thorny. Perhaps what irked Bohr was the way in which Pauli's early failures to derive the hydrogen molecule ion and the helium atom from Bohr's theory of the atom seriously undermined it. Pauli's later discovery of the fourth quantum number was one of the final pegs in its coffin. And his realization (as he tried to unravel the anomalous Zeeman effect) that all models of the atom with an inert core were wrong of course included Bohr's. Pauli's incessant criticisms too often annoyed Bohr—as dramatized in the spoof *Faust in Copenhagen* in which God stood for Bohr and Mephistopheles for Pauli. Indeed four years after Pauli's death, Bohr chose to mention his high regard for Stoner's work which he had, until then, kept to himself. In a shockingly unfair assessment of events some four decades earlier, he said: "Pauli was absolutely wonderful, but there was absolutely not a word that is new in the Pauli principle. This was all done by Stoner . . . one could have really called it the Stoner principle." (Interview with Bohr by T. S. Kuhn, *AHQP*, November 7, 1962, p. 10.) In fact, although Stoner was close to discovering the exclusion principle, he was unable to make the final leap. With a deeper understanding of the problem, Pauli did.

175 *"excluding* **Pauli himself"**: Speech by Panofsky, December 10, 1945, CERN Archive Collection, Document PLC Bi 264.

176 **"I feel, however, that I am European"**: Pauli to Casimir, October 11, 1945: *PLC3* [780].

176 **"the same about the spiritual situation"**: Pauli to Casimir, October 11, 1945: *PLC3* [780].

176 **"be brought under international control"**: Pauli to Klein, September 4, 1945: *PLC3* [767].

176 **"when those 'A-bombs' were dropped"**: Pauli to von Franz, May 17, 1951: *PLC4* [1239].

176 **Swiss citizen and a professor at the ETH:** Enz (1997), May, 5 1946, document II.134, which contains Roth's testimony on behalf of Pauli to Switzerland's Supervisor of Schools.

177 *"rotation are being tried"*: *P/J* [32P], October 28, 1946.

177 **Kepler's mandala is static and cannot rotate:** Pauli (1952), p. 234.

178 **the limitations of modern science:** Pauli (1952), p. 258.

178 **"system of resonators"**: Pauli (1952), p. 258.

178 **space and time were relative to God:** Pauli to Fierz, December 29, 1947: *PLC3* [926].

178 **"in the unconscious of modern man"**: *P/J* [33P], December 23, 1947.

178 **Jung was in the audience:** *P/J* [33P], December 23, 1947. Pauli gave two lectures—February 28 and March 6, 1948. For a summary see *P/J*, pp. 203–209.

179 **"a scientific theory of nature"**: *P/J*, p. 204.

179 **"and their archetypal foundations"**: Jung (1948), p. 473.

179 **"that amusing 'Pauli effect' "**: *P/J* [34P], June 16, 1948.

179 **"quantitative and figurative—i.e., symbolic sense"**: Pauli (1948a), p. 179.

179 **"concepts were still relatively undeveloped"**: Pauli (1948a), p. 179.

180 **" 'background physics' is of an archetypal nature"**: Pauli (1948a), p. 170.

180 **"into the world of physics"**: Pauli (1948a), p. 180.

180 **to translate the concept of spectral lines:** Pauli (1948a), p. 182.

180 **appears as two spectral lines:** Pauli (1948a), p. 183. Pauli was thinking of hydrogen. The hydrogen atom has one electron in its shell, so each of its spectral lines splits into two lines to give it a fine structure.

    Pauli first mentioned his interest in a neutral language in 1935. See *P/J* [13P], in which he looked into "the use of physical analogies to denote psychology facts in my dreams."

180 **the conscious as the mirror of the unconscious:** Pauli (1948a), p. 186.

180 **spectral lines on a photographic plate:** *P/J*, Appendix 7, p. 210. The editors identify this as a "Handwritten note from Pauli, undated." Actually, it was an attachment to Pauli's letter to Jung of December 23, 1953 (included in *P/J* [66P])—see *PLC5* [1695].

181 **a separation defined by 137:** Pauli (1948a), p. 191.

181 **belief that 137 was an archetypal number:** See Pauli (1948a), pp. 187 and 189.

182 **the whole process seems like nonsense:** See Jung (1930).

182 **"the divisibility of a quantity by *four*"**: The first quotation is from *P/J* [38P], June 4, 1950 and the second from *P/J* [23P], October 15, 1938.

182 **"main source of the feeling of harmony"**: *P/J* [23P], October 15, 1938.

182 **"And examines himself"**: Wilhelm (1923), p. 649.

Chapter 11 · Synchronicity

185a **clash of the four opposing concepts:** Pauli (1948a), p. 191. I have replaced Pauli's diagram with one from a letter he wrote to Jung two years later in 1950. The diagram is a visual representation of his text—see *P/J* [45P], November 24, 1950.

185 **"inherent in the distinction between subject and object"**: Bohr (1961), p. 91.

185 **"both negatively and positively (creatively)"**: Pauli (1948a), p. 192.

186 **an egg that then divides into two eggs:** Pauli (1948a), pp. 192–194.

187 **"symbolic description [of nature] par excellence"**: Pauli (1948a), p. 195.

187 **as well as the unconscious:** Pauli (1948a), p. 191.

188 **parapsychological phenomena as a medical student:** The term "synchronicity" appears for the first time in their correspondence in Pauli's letter to Jung of November 7, 1948, in which he wrote of their discussing "the 'synchronicity' of dreams and external circumstances" (*P/J* [35P]: November 7, 1948).

188 **"psychic and physical sequence of events come about":** *MDR*, p. 407.

188 **out-of-body occurrences and mental states:** Jung (1952a), p. 437.

189 **to Pauli about Dunne's clairvoyance:** *P/J* [8J], October 29, 1934. Dunne's book was a great success, especially the third edition which carried a note by Arthur Stanley Eddington in which he agreed that any theory of physics based solely upon cause and effect could not fully explain the world about us. See Dunne (2001), p. 132.

189 **"tentatively called the *synchronistic* principle":** Jung (1930), p. 56.

189 **"one of the best ideas":** Jung to Wylie, partial handwritten letter undated, in Philip Wylie Archives, Princeton University, quoted from Bair (2004), p. 551.

189 **"manifestations of the unconscious":** Jung (1930), p. 56.

190 **"preoccupation with psychic phenomena":** *P/J* [7P], October 26, 1934.

190 **"who looks at an infinite number of objects":** *P/J* [8J], October 29, 1934.

190 **"naturally regarded by us as superstition":** Jung to Jordan, November 10, 1934, in *CL*, vol. 1.

190 **"paying serious attention to such ideas":** *P/J* [36J], June 22, 1949.

191 **"the statistical method":** *P/J* [45P], November 24, 1950.

192 **"psychic state of the participants":** *P/J* [45P], November 24, 1950.

192 **But they existed nonetheless:** Jung (1952a), p. 424; and *P/J* [46J], November 30, 1950.

192 **"see any archetypal basis (or am I wrong there?)":** *P/J* [37P], June 28, 1949. Rhine's experimental results have been severely criticized. His experimental procedures lacked strict controls and several cases of falsification of records—in addition to just plain cheating—by his laboratory assistants, who often served as test subjects as well, have come to light.

192 **"psychologists of the unconscious *to this*":** Pauli to Fierz, March 20, 1950: *PCL4* [1091].

193 **constructing a scientific basis for astrology:** Jung (1952a), p. 475. Jung ended up omitting a great deal from his initial drafts. The manuscripts to Jung's "Synchronicity" paper can be studied at the Historisches Sammlungen, ETH Library, file Hs 1055:867, 1 and 2.

193 **there is no cause-and-effect connection:** Jung (1952a), p. 474.

193 **"a certain spiritual fertilization takes place":** Pauli to Fierz, November 26, 1949: *PLC3* [1058].

193 **tended to use synonymously with "simultaneity":** *P/J* [38P], June 4, 1950.

194 **"hypothesis of the collective unconscious":** Jung (1952a), pp. 439–440.

194 **given that the scarab is an ancient Egyptian symbol of rebirth:** Jung (1952a), pp. 426, 427, 438.

195 **"a direct connection with an archetype":** Jung (1952a), p. 481.

196 **"he is, like Merlin, in need of redemption":** Pauli to Emma Jung, in *P/J* [44P], November 16, 1950.

196 "That, I believe, is the myth of my life": Pauli to Jaffé, January 23, 1952: *PCL4* [1350].

196 "in order to determine the time": Pauli to Jung, February 2, 1951: *PLC4* [1200]. The set of dreams that Pauli enclosed with this letter was not included in the version published in *P/J* [50P]. It turned up in the correspondence of Jung's secretary, Aniela Jaffé, who often acted as a clearinghouse.

196 "not-yet-assimilated thoughts": *P/J* [39J], June 20, 1950.

197 worldview so deplored by Fludd: *P/J* [39J], June 20, 1950.

197 "the idealization of observation and definition respectively": Bohr (1961), pp. 54–55.

199 "no shortage of 'not-yet-assimilated thoughts' ": *P/J* [40P], June 23, 1950.

200 "with the aid of the new principle of synchronicity": *P/J* [45P], November 24, 1950.

200 a coincidence in time—total nonsense to a physicist: *P/J* [45P], November 24, 1950.

201 "half-life phenomenon of radium decay": ETH-Bibliothek Hs 1055:867, 1. Jung's other instances of physical phenomena on which synchronicity might be informative were: "The greatest density of water is at 4° c[entigrade]; The energy quanta; and the potential existence of the crystal lattice in the liquid that grows crystals."

201 the synchronistic phenomenon would immediately vanish: *P/J* [37P], June 28, 1949.

201 "a psychic state and a nonpsychic state": *P/J* [46J], November 30, 1950.

202 "has not lost its validity against the unconscious": *P/J* [46J], November 30, 1950.

202 devised with Pauli's help: Jung (1952a), p. 514.

202 "the psychology of the unconscious on the other hand": *P/J* [46J], November 30, 1950.

202 "the mathematical concept of probability": *P/J* [46J], November 30, 1950.

202 "the probability of psychic events": *P/J* [49J], January 13, 1951.

203 can do that is important: *P/J* [67J], October 10, 1955.

203 "avoided because of their bad name": Jung (1952a), p. 513.

203 "appreciate the psychological arguments": Jung (1952a), p. 514.

204 "in the human psyche": Pauli (1952), p. 221.

204 "a system of natural laws (that is, a scientific theory)": Pauli (1952), p. 220.

204 "constructing such a link": Pauli (1952), p. 220.

204 "distinct from the world of phenomena": Pauli (1952), p. 221.

204 "postulate of a cosmic order": Pauli (1952), p. 220.

204 "images with strong emotional content": Pauli (1952), p. 221.

204 "happiness that man feels in understanding" nature: Pauli (1952), p. 221.

205 **"emotional content stemming from the unconscious"**: Pauli (1952), p. 234.

205 **"a differentiation that can be traced throughout history"**: Pauli (1952), p. 258.

205 **"power of this number"**: Pauli (1952), p. 258; Fludd (1621), quoted from Pauli (1952), p. 273.

205 **"one can fall down on both sides"**: Pauli to Weisskopf, February 8, 1954: *PLC5* [1716].

205 **"a flight from the merely rational"**: Pauli (1955a), p. 147.

206 **"the redeeming experience of oneness"**: Pauli (1955a), p. 139.

206 **"concept of complementarity"**: Pauli (1952), p. 260.

206 **"it is impossible ever fully to understand the totality of nature"**: Pauli (1952), p. 259.

206 **"and [one that] can embrace them simultaneously"**: Pauli (1952), p. 259.

206 **"neither Pauli nor Jung needed much persuading"**: *P/J*, p. 81.

207 **"Dixi et salvavi animam meam! [I spoke and thus saved my soul]"**: Pauli to Fierz, December 25, 1954: *PLC* [1953].

Chapter 12 · Dreams of Primal Numbers

209 **"an urgent necessity for modern man"**: *P/J* [55P], February 27, 1952.

209 **Schopenhauer and Lao-tse on the philosophical side**: Pauli to Weisskopf, February 23, 1954: *PLC5* [1725].

209 **"strikes me as an untenable anthropomorphism"**: *P/J* [55P], May 17, 1952.

209 **"*Judeo-Christian monotheism is of no use to me*"**: Pauli to Jaffé, November 28, 1950: *PLC4* [1172].

209 **"akin to Buddhism"**: Jung (1951), p. 136.

210 **holes in a sea of negative-energy states**: Pauli to Fierz, June 2, 1949: *PLC3* [1029].

210 **"since it excludes the power of evil"**: Jung (1951), p. 41.

210 **"to the catastrophe that bears the name of Germany"**: Jung (1947), p. 170.

210 **and became his close friend**: Pauli to von Franz, end of December 1951: *PLC4* [1334].

211 **"unvarnished spectacle of divine savagery and ruthlessness"**: Jung (1952b), p. 366.

211 **in the form of the atomic bomb**: Jung (1952b), p. 351.

211 **often tormented him in his dreams**: *P/J* [58P], February 27, 1953.

211 **"*a new professorship*"**: *P/J* [58P], February 27, 1953.

211 **"something specific from the public"**: *P/J* [58P], February 27, 1953.

211 **also about psychology and even ethical problems**: *P/J* [58P], February 27, 1953.

212 **"So that was really tragic"**: Interview with Marie-Louise von Franz by Hein Stufkens and Philip Engelen, IKON-television, Küsnacht, November 1990.

212 **"exploited for both good and evil"**: P/J [58P], February 27, 1953.

212 **"it is also the same with me"**: P/J [64J], October 24, 1953.

212 **"a lot to tell the *strangers* about"**: P/J [59J], March 7, 1953.

212 **images from psychology by night**: P/J [60P], March 31, 1953.

212 **"*correspondentia* between psychological and physical fact"**: P/J [62P], May 27, 1953.

213 **"in geomancy and horoscopy"**: P/J [67J], October 10, 1955.

213 **"In my view both are true"**: P/J [67J], October 10, 1955.

213 **"successful in the field of archetypal ideas"**: P/J [64J], October 24, 1953.

213 **boat trips together on Lake Zürich**: See interview with von Franz by Gieser, excerpts in Gieser (2005), pp. 148–149.

214 **in a box in Pauli's office at the ETH**: See van Erkelens (1991).

214 **"on the *archetypal meaning of numbers*"**: Pauli to von Franz, March 9, 1952: *PLC4* [1383].

214 **"*further development* of Babylonian number mysticism"**: Pauli to von Franz, March 9, 1952: *PLC4* [1383].

214 **"to the concepts 'even' and 'odd' "**: Pauli to von Franz, March 9, 1952: *PLC4* [1383].

214 **"emerged such exact abstract concepts"**: Pauli to von Franz, March 9, 1952: *PLC4* [1383].

214 **"to find every pair of 'friendly numbers' "**: Pauli to von Franz, March 9, 1952: *PLC4* [1383].

215 **"a *psychological* problem connected with numbers"**: Pauli to von Franz, March 9, 1952: *PLC4* [1383].

215 **enable him to find all the friendly numbers**: Von Franz meant her book of 1970 to be the completion of Jung's work on the archetypal nature of numbers.

215 **He dedicated it to von Franz**: *The Piano Lesson* is in the Appendix to Pauli to von Franz, October 30, 1953: *PLC5* [1667].

216 **holding forth on psychology, physics, and biology**: Pauli saw a role for synchronicity in making Darwinian evolutionary theory more palatable. The complex end products of evolution were not just the result of directionless random mutations, with some selection and input from the environment, but "meaningful or purposeful coincidences of causally unconnected events." This was a third path in addition to those offered by Darwin and Lamarck (see *The Piano Lesson*). Pauli discussed his neo-Darwinism in lectures and with colleagues, especially his friend Max Delbrück who had moved from physics into biology and would be awarded a Nobel Prize in that field in 1969. In 1943 Delbrück had found results strongly suggesting that the environment had no effect on bacterial mutations. Most biologists interpreted this to mean that genetic changes were purely random. Further successes of molecular

biology were taken as meaning that our understanding of evolution could one day be reduced to the laws of quantum physics. This did not sit well with Pauli and no doubt was one of the reasons behind his neo-Darwinian position.

Delbrück was infuriated. He described Pauli as acting as if he were part of a "conspiracy of unemployed physicists against biology [and this is] simply stupid [and] not interesting at all" (Delbrück to Heisenberg, March 15, 1954: *PLC5* [1744]). See also Atmanspacher and Primas (2006) for a discussion of modern data that, to some degree, square with Pauli's neo-Darwinian view in that not all mutations are random. They argue that "Pauli's uneasiness with the straight Darwinian picture of biological evolution was fully justified" (p. 41).

216 the unity of wave and particle: See *The Piano Lesson* in Pauli to von Franz, October 30, 1953: *PLC5* [1667].

217 "get from Three to Four": *P/J* [64J], October 24, 1953.

217 "That was really the *main work*": Pauli to Fierz, October 3, 1951: *PLC4* [1286].

217 "The professor who shall calculate numerically": Pauli to von Franz, November 6, 1953: *PLC5* [1669].

217 *"what is still older is always the newer"*: Pauli to von Franz, November 6, 1953: *PLC5* [1669].

217 "believe that the full moon and the new moon": Pauli to von Franz, November 6, 1953: *PLC5* [1669].

217 Cambridge Platonist Thomas More: Fierz to Pauli, October 9, 1953: *PLC5* [1648].

218 "the dark half relapsed into the unconscious": Pauli to Fierz, October 11, 1953: *PLC5* [1651].

218 when these topics were hotly debated: Pauli to von Franz, November 6, 1953: *PLC5* [1669].

218 "God speaks to us *always* in riddles": Pauli to von Franz, November 6, 1953: *PLC5* [1669].

218 "until the summer of 1953": Pauli to von Franz, November 6, 1953: *PLC5* [1669].

218 Pauli's "active imagination": Pauli to von Franz, November 6, 1953: *PLC5* [1669].

219 In the youth phase of his life the archetypes are: Pauli to von Franz, November 6, 1953: *PLC5* [1669].

220 this woman is no longer his mother: Pauli to von Franz, November 6, 1953: *PLC5* [1669].

220 "anima-projections upon actual women": Pauli to von Franz, November 6, 1953: *PLC5* [1669].

220 "two aspects of the same problem": Pauli to von Franz, November 10–12, 1953: *PLC5* [1672].

221 **the number of signs in the zodiac:** Pauli to von Franz, November 10–12, 1953: *PLC5* [1672].

221 **"a problem with thorns and horns":** Pauli to von Franz, November 10–12, 1953: *PLC5* [1672].

221 **"every conceivable combination of chess":** Pauli to von Franz, November 10–12, 1953: *PLC5* [1672]. This dream, as well as Pauli's subsequent vision, is also in von Franz (1970), pp. 108–109, where she referred to Pauli as "a modern physicist."

221 **"such experiences with physics problems":** Pauli to von Franz, November 10–12, 1953: *PLC5* [1672]. This is characteristic of highly creative thinking— illumination after a period of unconscious thought. For further discussion see Miller (2000), chapter 9.

222 **" '*must be dynamically expressed*' ":** Pauli to von Franz, November 10–12, 1953: *PLC5* [1672].

222 **"6 lines and play of transformations":** Pauli to von Franz, November 10–12, 1953: *PLC5* [1672].

223 **'must either fly apart or flow into the other':** Fierz to Pauli, October 13/16, 1953: *PLC5* [1652].

223 **"then also is the 12 'incomplete' ":** Pauli to von Franz, November 10–12, 1953: *PLC5* [1672].

224 **"the exact *symmetrical* mental attitude of the pairs":** Pauli to von Franz, November 10–12, 1953: *PLC5* [1672].

224 **appealed to him "instinctively":** Pauli to von Franz, November 10–12, 1953: *PLC5* [1672].

224 **never discussed it in their letters:** Pauli's horoscope was deposited at the ETH (Wissenschaftshistorische Sammlung, Heisenberg's, Hs. 1056: 30880).

224 **"manifest itself both negatively and positively (creatively)":** Pauli (1948a), p. 192. See chapter 11.

224 **the instabilities of these boundaries between houses:** This description is based on a report Enz requested from an expert interpreter of horoscopes. The interpreter suggested an alternate and perhaps better astrological explanation of Pauli's sensitivity to the equinoxes in "The moon-lilith opposition on the equinoxal axis in his horoscope, 'Lilith' being the outer focus of the moon's elliptic orbit" (Enz [2002], pp. 464 and 498).

225 **"archetype of the quaternity":** *P/J* [66P], December 23, 1953.

225 **"what my physical dream symbolism is aiming at":** *P/J* [66P], December 23, 1953.

225 **"Certainly not I":** Pauli to Huxley, August 10, 1956: *PLC6* [2322].

225 **"beat me gently on the left ear":** Pauli to Sambursky, October 11, 1957: *PLC5*, p. xxxv.

226 **"but underdeveloped, even infantile":** Pauli to von Franz, April 15, 1951: *PLC4* [1209].

226 **became critical of him:** See Gieser (2005), pp. 148–149, 151. Gieser interviewed her in 1993.

226 **"that their relationship was tragic":** Gieser (2005), p. 151. From Gieser's interview of Meier in 1993.

226 **"Pauli himself never experienced any harm":** Quoted from Enz (2002), pp. 491–492.

226 **"a synchronistic phenomenon as conceived by Jung":** Quoted from Enz (2002), p. 150.

Chapter 13 · Second Intermezzo—Road to Yesterday

227 **"the shadow and my real father":** *P/J* [69P], October 23, 1956.

228 **"*expression with four quantities*":** *P/J* [69P], October 23, 1956. This was Pauli's dream of October 24, 1955.

228 **north and south, in a single object:** Jung (1951), pp. 287–346.

229 **"but not something to take seriously":** Interview with von Franz by Hein Stufkens and Philip Engelen, IKON-television, Künsnacht, November 1990.

229 **"To where is this journey":** Fierz to Pauli, January 22, 1956: *PLC6* [2209].

230 **The events took place during a trip to Hamburg:** The letters are: Pauli to Rosebaud, December 13, 1955: *PLC6* [2214]; Pauli to Fierz, March 2, 1956: *PLC6* [2253]; Pauli to Jung [69P], October 23, 1956.

230 **"this 'road to yesterday' ":** Pauli to Bohr, December 17, 1955: *PLC6* [2216].

230 **"poles in the pairs of opposites":** Pauli (1955a), p. 147.

230 **the phone in his hotel room rang:** Pauli to Fierz, March 2, 1956: *PLC6* [2253].

231 **"and she was in good health":** Pauli to Fierz, March 2, 1956: *PLC6* [2253].

231 **"National Socialism as a historical background":** *P/J* [69P], October 23, 1956.

231 **"but now it was very human":** *P/J* [69P], October 23, 1956.

231 **" 'What price glory' ":** Pauli to Rosebaud, December 13, 1955.

231 **"what should I say":** Pauli to Fierz, March 2, 1956.

232 **"Now it was as friends—at that time *not*":** Pauli to Rosebaud, December 13, 1955.

232 **"as regards my mental and spiritual well-being":** *P/J* [69P], October 23, 1956. This is from a dream Pauli had experienced, October 24, 1955.

232 **"symmetry between past and present":** Pauli to Fierz, March 2, 1956: *PLC6* [2253].

232 **"make a journey as I have made to Hamburg":** Pauli to Fierz, March 2, 1956: *PLC6* [2253].

Chapter 14 · Through the Looking Glass

233 **"This secret fills us with *apprehension*":** *P/J* [76P], August 5, 1957. This is from a dream Pauli had November 24, 1954.

233 **Chinese philosophy seeks to reconcile opposites:** *P/J* [59J], March 7, 1953.

234 **"and [his] fear of them":** *P/J* [76P], August 5, 1957. This is from a dream Pauli had November 24, 1954.

234 **"the whole point is about distinguishing between the two":** *P/J* [76P], August 5, 1957. This is from a dream Pauli had November 24, 1954.

234 **"left is the mirror image of the right":** Jung (1944), p. 172.

234 **"focusing on that particular subject":** *P/J* [76P], August 5, 1957. This is from a dream Pauli had November 24, 1954.

235 **"*C*, *P*, and *T* taken individually":** *P/J* [76P], August 5, 1957. This is from a dream Pauli had November 24, 1954.

236 **"so many stations are involved":** Pauli (1955b), p. 30. See Enz (2002), chapter 12, for details.

236 **"We are happy to inform you":** Enz (1994), p. 19. Frederick Reines and Clyde Cowan made the discovery at the Savannah River Nuclear Reactor. Both men were on leave from Los Alamos.

237 **"Everything comes to him who knows how to wait":** Enz (1994), p. 19.

237 **"ready to bet a very high sum":** Pauli to Weisskopf, January 17, 1957: *PLC7* [2445].

238 **"behaved irrationally for quite a while":** *P/J* [76P], August 5, 1957.

238 **"It would have resulted in a heavy loss":** Pauli to Weisskopf, January 27/28, 1957: *PLC7* [2476].

238 **"It is our sad duty to make known":** Pauli to Bohr, January 19, 1957: *PLC7* [2457].

239 **"does not conform to our thinking":** Pauli to Fierz, February 18, 1957: *PLC7* [2527].

239 **"is still very dominant":** Pauli to Kronig, April 5, 1955: *PLC6* [2061].

239 **"We have to be prepared for surprises":** Pauli to Telegdi, January 22, 1957: *PLC7* [2465].

239 **"but specifically with the beta-decay":** Pauli to Weisskopf, January 27/28, 1957: *PLC7* [2476].

239 **"an intelligent and beautiful Chinese young lady":** *P/J* [76P], August 5, 1957.

240 **"This particle neutrino . . . still persecutes me":** Pauli to Wu, January 19, 1957: *PLC7* [2460].

240 **"God *is* a weak left-hander after all":** *P/J* [76P], August 5, 1957.

240 **" 'Chinese revolution' in physics":** *P/J* [76P], August 5, 1957.

240 "acknowledging the nature of my 'mirror complex' ": *P/J* [76P], August 5, 1957.

241 "unconscious motives when one is involved in something": From an interview with Pauli by H. Bender, April 30, 1957; in Tournier to Bender, March 18, 1957: *PLC7* [2586].

241 "conventional objections to certain ideas": *P/J* [76P], August 5, 1957.

242 "Director *Spiegler* [Reflector], please": *P/J* [76P], August 5, 1957.

242 "relationship between physics and psychology is that of a mirror image": *P/J* [76P], August 5, 1957.

242 "partial mirror images": *P/J* [76P], August 5, 1957.

243 "The '*CPT* theorem' was on everybody's lips": *P/J* [76P], August 5, 1957.

243 "a fixed point around which all else turned": Interview with T. D. Lee by the author, Columbia University, April 23, 2008.

243 "will also prove a form of reflection": *P/J* [76P], August 5, 1957.

243 "It's the Pauli effect": Interview with T. D. Lee by the author, Columbia University, April 23, 2008.

244 Pauli and Jung often discussed the subject: *P/J* [56P], May 17, 1952.

244 "on the part of 'higher worlds' and the like": *P/J* [77J], August 1957.

245 "already had a scent of asymmetry": *P/J* [77J], August 1957.

245 "God and mankind, eternal and transient": *P/J* [77J], August 1957.

245 "psychological base of the UFO phenomenon": *P/J* [77J], August 1957.

245 newspapers piled high in his study: Bair (2004), p. 569.

245 "As I was walking up the hill from Zollikon station": *P/J* [56P], May 17, 1952.

246 "no reliable sightings": Knoll to Pauli, December 9, 1957, in *P/J,* Appendix 5.

246 "People think I am more stupid than I am": Jaffé to Pauli, December 29, 1957, in *P/J* [79].

246 "There are a great many things going on": Quoted from Bair (2004), p. 573.

246 "many long journeys": Jaffé to Pauli, December 29, 1957, in *P/J* [79].

Chapter 15 · The Mysterious Number 137

247 the mistake that He had made: Salam (1980), p. 351.

247 "consequences for the structure of matter in general": Born (1935), p. 539.

249 "I do not believe in the possible future of mysticism": Pauli to Hertha Pauli, October 11, 1957: *PLC7* [2707].

250 ratio of two prime numbers: A proper theory of quantum electrodynamics free of infinities is essential for any reliable determination of the fine structure constant (alpha). In 1949, Richard Feynman, Julian Schwinger, and Sin-Itoro Tomonaga succeeded in formulating such a theory, which had eluded Heisenberg and Pauli during the 1930s. The most accurate value of alpha to date was found by Gabrielse and co-workers at Harvard University (2008) and required the numerical calculation of over eight hundred Feynman diagrams (see Hanneke [2008]). Briefly, they worked out alpha indirectly by measuring the magnetic moment of the electron. Quantum electrodynamics relates this to alpha and the over eight hundred Feynman diagrams. From this they determined that $1/\alpha = 137.035999084$.

250 **"It is reasonable to inquire"**: Eddington (1968), p. 319. See Miller (2005) for more on Eddington.

251 **"What do you think of Eddington's latest article"**: Bohr to Pauli, January 8, 1929: *PLC1* [213].

251 **"on Eddington (??)"**: Pauli to Bohr, January 16, 1929: *PLC1* [214].

251 **"for romantic poets and not for physicists"**: Pauli to Klein, February 18, 1929: *PLC1* [216].

251 **"it makes no sense"**: Pauli to Sommerfeld, May 16, 1929: *PLC1* [225].

251 **" 'Atomistik' of the electric charge"**: Pauli to Heisenberg, February 7, 1934: *PLC1* [352].

251 **"a deep unification of foundations"**: Pauli (1933), p. 204. See Miller (1995) for more on Heisenberg's and Pauli's research on quantum electrodynamics during the 1930s.

251 **relativity theory which deals with the universe**: It is in the realms of imagination for the speed of light and also Planck's constant to change; but the fine structure constant can never change because it is found in nature (the spacing between the fine structure of spectral lines—as found in the laboratory—is determined by $1/137$).

This shows yet another meaning of the fine structure constant—the closeness of the bond between the worlds of quantum theory and of relativity that dramatically exhibits itself when scientists talk about moving between the everyday world and the worlds of relativity and quantum theory.

The only way to move back from the worlds of relativity and quantum theory to our daily world would be by making the speed of light infinite and Planck's constant zero. Increasing the speed of light to infinity would make the relativity of time disappear. Reducing Planck's constant to zero would do the same to the quantum qualities of the uncertainty relations and the wave-particle duality. We could then re-enter our everyday world where your time is the same as mine and things can be either particles or waves but not both. (In the quantum world, of course, electrons and light can be particles and waves at the same time.) But what physicists tend to forget in making these journeys between worlds is that the speed of light and Planck's constant are both in the bottom line of the fraction for the fine structure constant: $2\pi e^2/hc$. But the denominator in a fraction can never become zero because the result would be infinity, which is impossible in physics. Nor can we set the speed of light as infinite, which would make the fine structure constant zero, because it isn't, it's always $1/137$. Both the speed of light and Planck's constant have to be altered at once to ensure that the fine structure constant remains unchanged.

The fine structure constant, written by Niels Bohr on the blackboard in the physicist Edward M. Purcell's office at Harvard University, 1961. Purcell, Norman Ramsey (another physicist at Harvard), and Bohr were discussing the subtleties in moving from the quantum to the classical world, when Bohr commented, "People say that classical mechanics is the limit of quantum mechanics when h goes to zero." He shook his finger, walked to the blackboard and wrote the fine structure constant. As he underlined h three times, Bohr turned and said, "You see, h is in the denominator."

252 **"Everything will become beautiful":** Pauli to Heisenberg, April 17, 1934: *PLC2* [369].

252 **"I have been musing over the great question":** Pauli to Heisenberg, June 14, 1934: *PLC2* [373].

252 **"an interpretation of the numerical value":** Pauli (1934), p. 104.

252 **"The Mysterious Number 137":** Born (1935).

253 **"the central problem of natural philosophy":** Born (1935), p. 545.

253 **"main problems of theoretical physics":** Enz (1997), p. 194, as recalled by a former student, Wilhelm Frank.

253 **"a *magic number* that comes to us with no understanding by man":** Feynman (1985), p. 129.

255 **additional spectral lines—a fine structure:** As a result of Sommerfeld's discovery of the fine structure constant, the quantity 2.7×10^{-11} ergs from Bohr's original theory was re-expressed

$$2.7 \times 10^{-11} \text{ergs} = \frac{mc^2}{2} \alpha^2$$

where α is the fine structure constant. (The numerator and denominator of $\frac{2\pi^2 me^4}{h^2}$ are multiplied with 2 and c^2, and then the defining equation for the fine structure constant, $\alpha = \frac{2\pi e^2}{hc}$ is applied.) This equation reveals the intimate relationship in atomic physics between the fine structure constant, α, and the signature equation of relativity, $E = mc^2$. It also reveals the "naturalness" of the fine structure constant in the way in which it emerges in equations describing the properties of atoms.

255 **ways of doing this, all of which end up with 137, are:** See Barrow (2003), pp. 73–76, for more examples.

257 **seeking mystical truths and developing farfetched theories:** http://home .earthlink.net/~mrob/pub/math/numbers-4.html#cult.

257 **derive it from quantum electrodynamics:** Heisenberg to Bohr, January 10, 1935: *PLC2*, p. 366.

258 **prominent scholar of Jewish mysticism:** Gershom Scholem was the person who had doubts about whether to visit Jung after the war. See chapter 10.

258 **"the number associated with the Kabbalah":** Weisskopf (1992).

259 **number of the beast—666—as follows:** http://dgleahy.com/dgl/p22.html

260 **"Never before or afterward":** Heisenberg (1971), p. 233.

260 **"it's a long way to go":** Heisenberg (1971), p. 234.

260 **"conceptions of striking brilliance":** Rosenfeld (1967), pp. 118–119.

261 **"there since the creation of the world":** Quoted from Elisabeth Heisenberg (1984), p. 145.

261 **"theory of the smallest particles":** Pauli to Jaffé, January 5, 1958: *PLC8* [2825].

261 **"what I should write and calculate":** Pauli to Jaffé, January 5, 1958: *PLC8* [2825].

261 **"mirror complex":** Pauli to Jaffé, January 5, 1958: *PLC8* [2825].

261 **about which he was still exultant:** Pauli to Jaffé, January 5, 1958: *PLC8* [2825].

262 **"could almost feel the silence":** Interview with T. D. Lee by the author, Columbia University, April 23, 2008. At Pauli's request Lee scheduled the lecture at Columbia University. Lee told me that he personally had misgivings about it right from the start, but Pauli insisted.

262 **but it was clear that the passion had gone:** Communication from Eugen Merzbacher, who was there.

262 **the audience burst into applause:** Dyson to von Meyenn, December 20, 1958: *PLC8*, p. 872. Freeman J. Dyson was also there.

262 **"I was greatly saddened":** Yang to von Meyenn, August 27, 2002: *PLC8*, p. 871.

262 **"the death of a noble animal":** Communication from Jeremy Bernstein who recalled this comment by Dyson during Pauli's lecture.

263 **"something entirely new, in other words very 'crazy' ":** Pauli to Wu, November 17, 1958: *PLC8* [3111].

263 **"It's so totally stupid":** Pauli to Fierz, April 6, 1958: *PLC8* [2956].

263 **"the value reduced to 1/250":** Quoted from *PLC8*, p. 781.

263 **"Professor Heisenberg and his assistant":** Weisskopf to Pauli, March 7, 1958: *PLC8* [2912].

263 **"Only the technical details are missing":** Pauli to Gamow, March 1, 1958: *PLC8*: [2992].

264 Pauli's comment about Titian well in mind: Weisskopf to Pauli, March 7, 1958: *PLC8* [2912].

264 "Since the Heisenberg equation is supposed to describe": de-Shalit to Pauli, March 6, 1958: *PLC8* [2910].

264 "certainly he needs vacations": Pauli to de-Shalit, March 11, 1958: *PLC8* [2917].

264 "Super-Faust, Super-Einstein and Super-man Heisenberg": Pauli to Wu, mid-March 1958: *PLC8* [2926].

265 "how he runs after me": Pauli to Fierz, May 13, 1958: *PLC8* [2992].

265 "a substitute for fundamental ideas": Quoted from Cassidy (1992), p. 537.

265 "don't laugh in the wrong place": Quoted from *PLC8*, p. 1218.

265 "Things have changed too much": Heisenberg (1971), p. 236.

265 "the sober American pragmatists": Heisenberg (1971), p. 234.

265 "his belief that that was possible": Franca Pauli to Mrs. Niels Bohr, in Pais (2000), p. 252.

266 "a classicist and no revolutionary": Interview by Jagdish Mehra, February 1958, in Mehra and Rechenberg (1982), p. xxiv.

266 "a certain defensive wariness": Glauber (2001).

267 "And how is your dear mother": I thank Jeremy Bernstein for this story.

268 "better than your astronomical work": Hoyle (1994), p. 310.

269 "wet after-session": Schucking (2001), p. 47.

269 "It's 137": Enz (2002), p. 533.

270 sparked his greatest discoveries: "Theoretical Physics of the ETH to Heisenberg," December 16, 1958: *PLC8* [3130].

270 they were too busy to attend: Elisabeth Heisenberg to Franca Pauli, dated as "End of 1958": *PLC8* [3134].

270 "I thought quite unreasonably": Heisenberg (1971), p. 235.

270 "remain my 'better English' ": CatholicAuthors.com.

270 "married her translator": Quoted from *PLC7*, p. 569.

271 "*Pauli's own words*": Enz to Franca Pauli, May 14, 1959: *PLC8* [3135].

271 "an article connected with the subject": Franca Pauli to Abdus Salam, January 4, 1962: *PLC8* [3139].

271 "I would not repeat now": Salam to Franca Pauli, January 9, 1962: *PLC8* [3140].

271 "could not characterize Pauli better in so few words": Franca Pauli to Salam, January 13, 1962: *PLC8* [3141].

Epilogue

274 preparing them for publication: See www.philemonfoundation.org.

274 free of the infinities that had rendered invalid: Bohr's theory of the atom

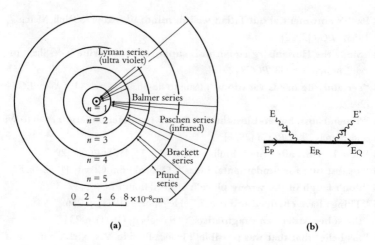

(a)   **(b)**

produces the visual image for light interacting with a hydrogen atom as shown in
Figure (a). Bohr adapted equations from how the earth goes around the sun to the
world of the atom. The lines indicate the spectral lines that emerge when the electron in the atom makes a quantum jump from a higher to a lower orbit.

The Feynman diagram in Figure (b) shows the same processes depicted in
Figure (a)—a hydrogen atom being struck by light. The line E_p is the state of the
atom before it is struck by a light quantum (the wavy line E). The line E_R shows all
the higher energy states of the atom which the electron can be driven up to after
having been struck by light. E' represents the light quantum that the atom emits
when its electron drops back to its ground state; it is the light split by a spectrom-
eter into spectral lines.

The Feynman diagram is strikingly different from the image produced in
Bohr's theory of the atom. We would not know how to draw it without the proper
equations—the equations of Feynman's theory for how light interacts with electrons, the new quantum electrodynamics free of infinities.

275 a "sentimental painting": Quoted from Enz (2002), p. 444.

Bibliography

Works cited frequently in the notes are identified by the following abbreviations:

AHQP   Archive for History of Quantum Physics. Office of History of Science and Technology, University of California, Berkeley.

CL   *C. G. Jung Letters,* selected and edited by Gerhard Adler in collaboration with Aniela Jaffé, translated by R.F.C. Hull, 2 vols (vol. 1: 1906–1950; vol. 2, 1951–1961), Bollingen Series 95.1 and 95.2. Princeton: Princeton University Press, 1973 and 1991.

CSP   Kronig, Ralph, and Victor Weisskopf, eds. 1964. *Collected Scientific Papers by Wolfgang Pauli.* New York: Wiley.

CW   *The Collected Works of C. G. Jung,* Sir Herbert Reid, Michael Fordham, and Gerhard Adler (eds.), William McGuire (executive ed.), R.F.C. Hull (trans.), 20 vols.; Bollingen Series 20 (Princeton: Princeton University Press).

F/J   *The Freud/Jung Letters: The Correspondence Between Sigmund Freud and C. G. Jung.* Princeton: Princeton University Press, 1974. W. McGuire (ed.), R. Manheim and R.F.C. Hill (trans.).

MDR   *Memories, Dreams, Reflections* by C. G. Jung, recorded and edited by Aniela Jaffé, translated from the German by Richard and Clara Winston. (London: Fontana Press, 1995).

P/J   *Atom and Archetype: The Pauli/Jung Letters, 1932–1958.* Princeton: Princeton University Press, 2001. Charles A. Meier (ed.), with the assistance of C. P. Enz and M. Fierz, D. Roscoe (trans.), with an introductory essay by B. Zabriskie. Originally published as C. A. Meier (ed.) with assistance from C. P. Enz and M. Fierz, *Wolfgang Pauli und C. G. Jung: Ein Briefwechsel, 1932–1958* (Berlin: Springer-Verlag, 1992).

PLC1   *Wolfgang Pauli: Wissenschaftlicher Briefwechsel mit Bohr, Einstein, Heisenberg u.a. Band I, 1919–1929* [Scientific Correspondence with Bohr, Einstein, Heisenberg, a.o. Volume I: 1919–1929]. A. Hermann, K. von Meyenn, V. F. Weisskopf (eds.). Berlin: Springer-Verlag, 1979.

PLC2   *Wolfgang Pauli: Wissenschaftlicher Briefwechsel mit Bohr, Einstein, Heisenberg u.a. Band II, 1930–1939* [Scientific Correspondence with Bohr, Einstein, Heisenberg, a.o. Volume II: 1930–1939]. K. von Meyenn (ed.), with the cooperation of A. Hermann, V.F. Weisskopf. Berlin: Springer-Verlag, 1985.

PLC3   *Wolfgang Pauli: Wissenschaftlicher Briefwechsel mit Bohr, Einstein, Heisenberg u.a. Band III, 1940-1949* [Scientific Correspondence with Bohr, Einstein, Heisenberg, a.o. Volume III: 1940-1949]. K. von Meyenn (ed.). Berlin: Springer-Verlag, 1993.

PLC4   *Wolfgang Pauli: Wissenschaftlicher Briefwechsel mit Bohr, Einstein, Heisenberg u.a. Band IV, Teil I, 1950–1952* [Scientific Correspondence with Bohr, Einstein, Heisenberg, a.o. Volume IV, Part I: 1950-1952]. K. von Meyenn (ed.). Berlin: Springer-Verlag, 1996.

PLC5   *Wolfgang Pauli: Wissenschaftlicher Briefwechsel mit Bohr, Einstein, Heisenberg u.a. Band IV, Teil II, 1953–1954* [Scientific Correspondence with Bohr, Einstein, Heisenberg, a.o. Volume IV, Part II: 1953-1954]. K. von Meyenn (ed.). Berlin: Springer-Verlag, 1999.

PLC6   *Wolfgang Pauli: Wissenschaftlicher Briefwechsel mit Bohr, Einstein, Heisenberg u.a. Band IV, Teil III, 1955–1956* [Scientific Correspondence with Bohr, Einstein, Heisenberg, a.o. Volume IV, Part III: 1955–1956]. K. von Meyenn (ed.). Berlin: Springer-Verlag, 2001.

PLC7   *Wolfgang Pauli: Wissenschaftlicher Briefwechsel mit Bohr, Einstein, Heisenberg u.a. Band IV, Teil IV, A: 1957* [Scientific Correspondence with Bohr, Einstein, Heisenberg, a.o. Volume IV, Part IV, A: 1957]. K. von Meyenn (ed.). Berlin: Springer-Verlag, 2005.

PLC8   *Wolfgang Pauli: Wissenschaftlicher Briefwechsel mit Bohr, Einstein, Heisenberg u.a. Band IV, Teil IV, B: 1958* [Scientific Correspondence with Bohr, Einstein, Heisenberg, a.o. Volume IV, Part IV, B: 1958]. K. von Meyenn (ed.). Berlin: Springer-Verlag, 2005.

WPP   *Wolfgang Pauli: Writings on Physics and Philosophy*. C. P. Enz and K. von Meyenn (eds.) and R. Schlapp (trans). Berlin: Springer-Verlag, 1994.

Selected Bibliography

To set writings in their historic context I will refer to a book or article with its earliest publication date.

Aaron, Daniel. 2001. "The Great Good Place." *Harvard Magazine*. September–October 2001.

Andrade, E.N.C. (1926). *The Structure of the Atom*. 3rd ed. New York: Harcourt, Brace & Co.

Atmanspacher, Harald, and Hans Primas. 2006. "Pauli's Ideas on Mind and Matter in the Context of Contemporary Science." *Journal for Consciousness Studies* 13: 5–50.

Bair, Deirdre. 2004. *Jung: A Biography*. New York: Little Brown.

Barrow, John. 2003. *The Constants of Nature: From Alpha to Omega*. London: Vintage.

Bernstein, Jeremy. 1997. "Heaven's Net: The Meeting of John Donne and Johannes Kepler," *American Scholar* 66, Issue 2: 175–195.

Bohr, Niels. 1932. "Chemistry and the Quantum Theory of Atomic Constitution." *Journal of the Chemical Society*. 349–384. This is Bohr's Faraday Lecture, delivered May 8, 1930.

Bohr, Niels. 1961. "The Quantum Postulate and the Recent Development of Atomic Theory." *Atomic Theory and the Description of Nature*. Cambridge: Cambridge University Press, 1961: 52–91. This is a published version of Bohr's lecture delivered September 16, 1927, to the International Congress of Physics, Como, Italy.

Born, Max. 1923. "Quantentheorie und Störungsrechnung." *Die Naturwissenschaften* 27: 537–550.

———. 1935. "The Mysterious Number 137," *Proceedings of the Indian Academy of Sciences*, Sec. A, 2: 533–561.

———. 1971. *The Born-Einstein Letters: Correspondence between Albert Einstein and Max and Hedwig Born from 1916 to 1955*. Translated by Irene Born. New York: MacMillan.

———. 1975. *My Life and Recollections as a Nobel Laureate*. New York: Scribner.

Brewster, David. 1855. *The Life of Sir Isaac Newton: The great philosopher*. Revised and edited by W. T. Lynn. London: Gall & Inglis.

Burrau, Øvind. 1926/1927. "Berechnung des Energiewertes des Wasserstoffmolekel-Ions (H_2^+) im Normalzustand" [Calculation of the Energy Values of the Hydrogenmoluclue ion (H_2^+) in the Normal State]. Kongelige Danske Videnskabernes Selskab Mathem. Fysiske Meddelelser, Volume 7, Number 14.

Caspar, Max. 1948. *Kepler*. Translated by C. Doris Hellman from the German edition of 1948. New York: Dover.

Cassidy, David. 1992. *Uncertainty: The life and science of Werner Heisenberg*. New York: W. H. Freeman.

Cline, Barbara L. 1987. *Men Who Made a New Physics: Physicists and the quantum theory*. Chicago: University of Chicago.

Dan, Joseph. 2006. *Kabbalah: A very short introduction*. Oxford: Oxford University Press.

Dirac, P.A.M. 1926. "On the Theory of Quantum Mechanics," *Proceedings of the Royal Society (London) A*, 112: 661–677.

———. 1927. "The Quantum Theory of the Emission and Absorption of Radiation," *Proceedings of the Royal Society (London) A*, 114: 243–265.

Drab, Sanford. 2005. "Jung's Kabbalistic Visions," *Journal of Jungian Theory and Practice* 7: 33–54.

Dunne, John William. 2001. *An Experiment with Time*. Charlottesville, VA: Hampton Roads Publishing Company, Inc.; first published 1927.

Eckert, M., and K. Märker, eds. 2000. *Arnold Sommerfeld. Wissenschaftlicher Briefwechsel*. Berlin: GNT-Verlag.

Eddington, Arthur Stanley. 1968. *The Nature of the Physical World*. Ann Arbor: University of Michigan Press.

Einstein, Albert. 1918. "Principles of Research." *Essays in Science*. New York: Philosophical Library: 1–5.

———. 1922. [Review of] W. Pauli: *Relativitätstheorie, Die Naturwissenschaften* 10: 184.

Enz, Charles P. 1994. "A Biographical Introduction." In *WPP*, pp. 13–26.

———. 1997. With B. Glaus and G. Oberkofler, eds. *Wolfgang Pauli und sein Wirken an der ETH Zürich*. Zürich: vdf Hochschulverlag ETH.

———. 2002. *No Time to Be Brief: A scientific biography of Wolfgang Pauli*. Oxford: Oxford University Press.

———. 2005. *Pauli hat gesagt: Eine Biographie des Nobelpreisträgers Wolfgang Pauli, 1900–1958*. Zürich: Verlag Neue Züricher Zeitung.

Feynman, Richard. 1985. *Quantum Electrodynamics: The strange story of light and matter*. Princeton: Princeton University Press.

Fierz, Markus, and Victor F. Weisskopf, eds. (1960). *Theoretical Physics in the Twentieth Century: A memorial volume to Wolfgang Pauli*. New York: Interscience Publishers.

Fludd, Robert. 1621. *Demonstratio quaedam analytica (Discursus analysticus)*. Frankfurt: Johann-Theodor de Bry.

Franz, Marie-Louise von. 1970. *Number and Time: Reflections leading toward a unification of depth psychology and physics*. Translated by Andrea Dykes. Evanston, IL: Northwestern University Press, 1974. Originally published as *Zahl und Zeit: Psychologische Überlegungen zu einer Annäherung von Tiefenpsychologie und Physik* (Stuttgart: Ernst Klett Verlag).

———. 1972. *C. G. Jung: His myth in our time*. Translated by William H. Kennedy. Toronto: Inner City Books.

———. 1972. *C. G. Jung: His myth in our time*. Toronto: University of Toronto Press. Translated by William H. Kennedy.

Gamow, George. 1985. *Thirty Years That Shook Physics: The story of quantum theory.* New York: Dover.

Gay, Peter. 1989. *Freud: A life in our time.* New York: Doubleday.

Gieser, Suzanne. 2005. *The Innermost Kernel: Depth psychology and quantum physics. Wolfgang Pauli's dialogue with C. G. Jung.* Berlin, Springer-Verlag.

Glauber, Roy. 2001. "A Remembrance of Pauli in 1950." *Physics Today* 49: 43.

Hanneke, D., S. Fogwell, and G. Gabrielse. 2008. "New Measurement of the Electron Magnetic Moment and the Fine Structure Constant," *Physical Review Letters* 100, 1207801.

Hayman, Ronald. 1999. *A Life of Jung.* London: Bloomsbury Publishing Plc.

Heilbron, John. 1983. "The Origins of the Exclusion Principle." *Historical Studies in the Physical Sciences* 13: 261–310.

Heisenberg, Elisabeth. 1984. *Inner Exile: Recollections of a life with Werner Heisenberg.* With an Introduction by Victor Weisskopf. Boston: Birkhäuser. Translated by S. Cappellari and C. Morris from *Das politische Leben eines Unpolitischen—Errinerungen an Werner Heisenberg* (Munich: Piper Verlag, 1980).

Heisenberg, Werner. 1926. *"Mehrkörperproblem und Resonzanz in der Quantenmechanik," Zeitschrift für Physik* 38: 411–426.

———. 1960. *"Errinerungen an die Zeit der Entwicklung der Quantenmechanik."* In Fierz, Marcus, and Victor F. Weisskopf, eds. *Theoretical Physics in the Twentieth Century: A memorial volume to Wolfgang Pauli.* New York: Interscience Publishers, pp. 40–47.

———. 1971. *Physics and Beyond: Encounters and conversations.* Translated by Arnold J. Pomerans. New York: Harper & Row.

Heisenberg, Werner, and Pascual Jordan. 1926, "Anwendung der Quantenmechanik auf das Problem der anomalen Zeemaneffekte," *Zeitschrift für Physik* 37: 263–267.

Heisenberg, Werner, and Wolfgang Pauli. 1929. "Zur Quantendynamik der Wellenfelder." *Zeitschrift für Physik* 56: 1–61.

———. 1930. "Zur Quantentheorie der Wellenfelder II." *Zeitschrift für Physik* 59: 168–190.

Hendry, John. 1984. *The Creation of Quantum Mechanics and the Bohr-Pauli Dialogue.* Dordrecht: Reidel.

Hermann, Armin, ed. 1968. *Albert Einstein / Arnold Sommerfeld Briefwechsel.* Basel and Stuttgart: Schwabe.

Hermann, Armin. 1979. "Die Funktion und Bedeutung von Briefen." *PLC1,* pp. XI–XLVII.

Holton, Gerald. 1973. "Johannes Kepler's Universe: Its Physics and Metaphysics." In G. Holton, *Thematic Origins of Scientific Thought: Kepler to Einstein.* Cambridge, MA: Harvard University Press, pp. 69–90.

Hoyer, U., ed. (1981). *Niels Bohr, Collected Works.* Amsterdam: North Holland.

Hoyle, Fred. 1994. *Home Is Where the Wind Blows: Chapters from a cosmologist's life.* Mill Valley, CA: University Science Books.

Jordan, Pascual. 1971. *Begegnungen.* Oldenburg-Hamburg: Gerhard Stallung Verlag.

Jaffé, Aniela. 1971. *From the Life and Work of C. G. Jung*. London: Hodder & Stoughton.

Jammer, Max. 1966. *The Conceptual Development of Quantum Mechanics*. New York: McGraw-Hill.

Jung, Carl. 1911/12. "*Symbols of Transformation*." In *CW5*.

———. 1916. "*Transzendente Funktion*" [The Transcendent Function]." Jung did not publish this manuscript for forty years. See *CW8*, pp. 67–91, where on pp. 67–68 is Jung's "Prefatory Note," written in 1958.

———. 1918. "The Role of the Unconscious." In *CW10*, pp. 3–28.

———. 1921 *Psychological Types: The psychology of individuation*. Tranlsated by H. Godwin Baynes. London: Kegan, Paul, Trench, Trubner & Co.

———. 1930. "Richard Wilhelm: In memoriam." Originally delivered at a memorial service for Wilhelm in Munich, May, 1930. In *CW15*, pp. 53–62.

———. 1934. "The State of Psychotherapy." In *CW10*, pp. 157–178.

———. 1935. "Tavistock Lectures." In *CW18*, pp. 5–182. This is a series of lectures Jung delivered at the Insitute of Medical Psychology (Tavistock Clinic) in London, September 30–October 4, 1935.

———. 1936a. "Psychological Factors Determining Human Behaviour." In *CW8*, pp. 114–125.

———. 1936b. "Dream Symbols of the Individuation Process." Seminar held at Bailey Island, Maine, September 20–25, 1936 (unpublished notes).

———. 1936c. "Wotan." In *CW10*, pp. 179–193.

———. 1937a. "Dream Symbols of the Individuation Process," October 16–18 and 25, 1937, New York City (unpublished).

———. 1937b. "Psychology and Religion," Terry Lectures, October 20, 21, 22, and 24, 1937, Yale University. Published with revisions and augmentations in *CW11*, pp. 3–105.

———. 1944. "Individual Dream Symbolism in Relation to Alchemy: A Study of the Unconscious at Work in Dreams." In *CW12*, pp. 39–223. This is a translation of a portion of Jung's, *Psychologie und Alchemie* (Zürich: Rascher Verlag, 1944).

———. 1947. "On the Nature of the Psyche." In *CW8*, pp. 159–234.

———. 1948. "Address on the Occasion of the Founding of the Jung Institute." In *CW18*, pp. 471–476.

———. 1949. "Foreword" to Richard Wilhelm, *I Ching or Book of Changes*. London: Routledge & Kegan Paul Ltd. Translated by C.F. Baynes from the German edition of 1923, pp. xxi-xxxix.

———. 1951. "*Aion: Researches into the Phenomenology of the Self*." In *CW9, Part 2*. Translated from the first part of *Aion: Untersuchungen zur Symbolgeschichte* (Zürich: Rascher Verlag, 1951).

———. 1952a. "Synchronicity: an acausal connecting principle." In *The Interpretation of Nature and the Psyche*. New York: Bollingen Series LI, Pantheon Books, 1955. Translation by Priscilla Silz from the German edition of 1952. Reprinted in *CW8*, pp. 417–531. All references are to *CW8*.

————. 1952b. *Answer to Job*. In *CW11*, pp. 355–470. Translated from *Antwort auf Hiob* (Zürich: Rascher Verlag, 1952).

————. 1958a. "Preface to de Laszlo: 'Psyche and Symbol'." In *CW18*, pp. 537–542. This volume is a selection of Jung's writings by Violet S. de Laszlo (New York: Anchor Books).

————. 1958b. "Flying Saucers: A modern myth of things seen in the skies." In *CW10*, pp. 307–433. First published as *Ein moderner Mythus: Von Dingen, die am Himmel gesehen werder* (Zürich: Rascher Verlag).

————. 1959. *Über Gefühle und den Schatten* [On Sensation and Shadows]. Zürich: Walter. Textbook and CDs for the lecture delivered at a private event, June 27, 1959.

Keynes, John Maynard. 1947. "Newton the Man." In Royal Society, *Newton Tercentenary Celebrations*. Cambridge: Cambridge University Press, pp. 27–34.

Koestler, Arthur. 1964. *The Sleepwalkers: A History of Man's Changing Vision of the Universe*. London: Penguin.

Kragh, Helge. 2003. "Magic Number: A Partial History of the Fine-Structure Constant," Archive for History of Exact Sciences, Volume 57, pp. 395–431.

Kramers, Hendrik, with H. Holst. 1923. *The Atom and the Bohr Theory of Its Structure* Translated by R. B. and R. T. Lindsay. London: Gyldendal.

Kronig, Ralph. 1960. "The Turning Point." In M. Fierz and V. F. Weisskopf, eds. *Theoretical Physics in the Twentieth Century: A memorial volume to wolfgang/Pauli*. New York: Interscience, pp. 5-39.

Kuhn, Thomas S. 1957. *The Copernican Revolution: Planetary astronomy and the development of Western thought*. New York: Vintage Books.

Lindorff, David. 2004. With a Foreword by Markus Fierz. *Pauli and Jung: The meeting of two great minds* Wheaton, IL: Quest Books.

Mach, Ernst. 1910. "The Guiding Principles of My Scientific Theory of Knowledge." Translated by Ann Toulmin, in Stephen Toulmin, ed. *Physical Reality* (New York: Harper Torchbooks, 1970), pp. 28–43.

Massimi, Michela. 2005. *Pauli's Exclusion Principle: The origin and validation of a scientific principle*. Cambridge, UK: Cambridge University Press.

Mehra, Jagdish, and Helmut Rechenberg. 1982. *The Historical Development of Quantum Theory: Volume 1, Parts 1 and 2, The Quantum Theory of Planck, Einstein, Bohr and Sommerfeld: Its Foundation and the Rise of Its Difficulties 1900–1925*. Berlin: Springer-Verlag.

Meyenn, Karl von. 1980, 1981. "Pauli's Weg zum Ausschliessungsprinzip. Teil I und II," *Physickalisches Blätter* 36: 293–98 and 37: 13–19.

————. 1999. "*Pauli's philosophische Auffassungen*" [Pauli's philosophical conceptions]. In *PLC5*, pp. vi–xxxv.

————. 2007. "Markus Eduard Fierz (1912–2006)," *Mind and Matter* 5: 241–267.

Meyenn, Karl von, and Schucking, Engelbert. 2001. "Wolfgang Pauli." *Physics Today*, February, pp. 43–48.

Miller, Arthur I. 1995. *Early Quantum Electrodynamics*. Cambridge, UK: Cambridge University Press.

————. 2000. *Insights of Genius: Imagery and creativity in science and art.* Cambridge, MA: MIT Press.

————. 2005. *Empire of the Stars: Friendship, obsession and betrayal in the quest for black holes.* London: Abacus.

Pais, Abraham. 2000. *The Genius of Science: A portrait gallery.* Oxford: Oxford University Press.

————. 2006. *J. Robert Oppenheimer: A life.* With supplemental material by Robert P. Crease. Oxford: Oxford University Press.

Pauli, Hertha. 1970. *Der Riß der Zeit geht durch mein Herz* [The Rift of Time Goes Through My Heart]. Vienna: Zsolnay.

Pauli, Wolfgang. 1919. "Über die Energiekomponenten des Gravitationsfeldes" [On the Energy Components of the Gravitational Field], *Physikalische Zeitschrift* 20: 25–27. Received September 9, 1918.

————. 1925. "*Über den Zusammenhang des Abschlusses der Elektrontruppen im Atom mit der Komplexstruktur der Speketren*" [The Connection of the Closings of the Electron Groups in the Atom with the Complex Structure of Spectra], *Zeitschrift für Physik* 31: 765–783.

————. 1927. "*Über Gasentartung und Paramagnetismus*" [Degenerate Gases and Paramagnetism], *Zeitschrift für Physik* 41: 81–102.

————. 1933. *General Principles of Quantum Mechanics.* Berlin: Springer-Verlag, 1980. Translated by P. Achuthan and K. Venkatesan with additions from the original German edition, *Die allgemeinen Prinzipien der Wellenmechanik. Handbuch der Physik, herausgegeben von Geiger und Scheel. Two Volumes* (Berlin: Julius Springer, 1933).

————. 1934. "Space, Time and Causality in Modern Physics." In *WPP*, pp. 95–105. This is an expanded version of a lecture Pauli gave to the Philosophical Society in Zürich in November 1934.

————. 1940. "The Connection Between Spin and Statistics," *Physical Review* 58: 716–722.

————. 1946. "Remarks on the History of the Exclusion Principle." In *WPP*, pp. 165–181. This is Pauli's Nobel Prize lecture delivered at Stockholm on December 13, 1946. Quotations are from *WPP*.

————. 1948a. "Modern Examples of 'Background Physics'." In *P/J*, pp. 179–196.

————. 1948b. "Sommerfeld's Contributions to the Quantum Theory," *Die Naturwissenschaften* 35: 129. Reprinted in *WPP*. All references are to *WPP*.

————. 1952. "The Influence of Archetypal Ideas on the Scientific Theories of Kepler." In *The Interpretation of Nature and the Psyche.* Translated by Priscilla Silz. (New York: Bollingen Series LI, Pantheon Books, 1955). Reprinted in *WPP*, pp. 218–279. All references are to *WPP*.

————. 1955a. "Science and Western Thought." In *WPP*, pp. 137–148. This is a lecture Pauli delivered at the International Congress of Scholars, Mainz, May 18, 1955, and Hamburg, November 29, 1955.

————. 1955b. "Exclusion Principle, Lorentz Group and Reflection of Space-Time and Charge." In W. Pauli, ed., with the assistance of L. Rosenfeld and V. Weisskopf, *Niels Bohr and the Development of Physics: Essays dedicated to Niels Bohr on the occasion of his seventieth birthday.* London: Pergamon Press, pp. 30–51.

Peierls, Rudolf. 1960. "Wolfgang Ernst Pauli." In *Biographical Memoirs of Fellows of the Royal Society* 5: 174–192.

————. 1985. *Bird of Passage: Recollections of a physicist.* Princeton: Princeton University Press.

Powers, Thomas. 2000. *Heisenberg's War: The secret history of the German atomic bomb.* New York: Da Capo Press.

Rhine, Joseph B. 1935. *Extra-Sensory Perception.* London: Faber & Faber.

Roazen, Paul. 1975. *Freud and His Followers.* New York: Random House.

Rosenfeld, Léon. 1967. "Niels Bohr in the Thirties." In S. Rozental, ed., *Niels Bohr: His Life and Work as Seen by His Friends and Colleagues.* New York: Interscience, pp. 114–136.

Salam, Abdus. 1980. "Gauge Unification of Fundamental Forces." In *Reviews of Modern Physics* 52: 525–538.

Sauvage, Marcel. 1949. *Les mémoires de Joséphine Baker.* Paris, Éditions Correa.

Scholem, Gershom. 1941. *Modern Trends in Jewish Mysticism.* Jerusalem: Schocher Publishing House.

Schrödinger, Erwin. 1926. "Über der Verhältnis der Heisenberg-Born-Jordanschen Quantenmechanik zu der meinen" [On the Relation of the Heisenberg-Born-Jordan Quantum Mechanics to Mine], *Annalen der Physik* 79: 734–756.

Schucking, Engelbert. 2001. *Physics Today,* February: 46–47.

Segré, Gino. 2008. *Faust in Copenhagen: A Struggle for the Soul of Physics.* London: Penguin.

Silberer, Herbert. 1914. *Problems of Mysticism and Its Symbolism.* Translated by Smith Ely Smith (New York: Kessinger, 2006).

Sommerfeld, Arnold. 1923. *Atomic Structure and Spectral Lines* Translated from the third German edition by Henry L. Brose (London: Methuen).

————. (1927), "Über kosmische Strahlung" [On Cosmic Radiation], *Südd. Monatshefte* 24: 195–198.

————. 2004. *Wissenschaftliche Briefwechsel. Band 2: 1919–1951.* Edited by M. Eckert and K. Märker. Berlin, Diepholz, Munich: Deutsches Museum, Verlag für Geschichte der Naturwissenschaften und der Technik.

Telegdi, Valentine L. 1987. "Parity Violation," In M.G. Doncel, A. Hermann, L. Michel, and A. Pais, eds., *Symmetries in Physics (1600-1980): 1st International Meeting on the History of Scientific Ideas, Sant Feliu de Guixols, Catalonia, Spain. September 20-26, 1983* (Barcelona: Seminari d'Història de les Ciècies Universitat Autònoma de Barcelona), pp. 432–471.

Van Erkelens, Herbert. 1991. "Wolfgang Pauli's Dialogue with the Spirit of Matter," *Psychological Perspectives* 24.

Van der Waerden, Bartel Leendert. 1954. *Science Awakening.* Translated by Arnold Dresden (Gorningen: Noordoff).

Weisskopf, Victor. 1989. *The Privilege of Being a Physicist.* New York: W.H. Freeman.

———. 1992. *The Joy of Insight: Passions of a Physicist.* New York: Basic Books.

Wentzel, Gregor. 1960. "Quantum Theory of Fields (until 1947)." In M. Fierz and V. F. Weisskopf, eds., *Theoretical Physics in the Twentieth Century: A memorial volume to Wolfgang Pauli* (New York: Wiley), pp. 48–77.

Westman, Robert. 1984. "Nature, Art and Psyche: Jung, Pauli, and the Kepler-Fludd Polemic." in Brian Vickers, ed., *Occult and Scientific Mentalities in the Renaissance* (Cambridge: Cambridge University Press), pp. 177–229.

Wilhelm, Richard. 1923. *I Ching: or Book of Changes.* London: Routledge & Kegan Paul Ltd. Translated by C.F. Baynes from the German edition of 1923 with a Foreword by C. G. Jung. All references are to the English-language version.

———. 1929. *The Secret of the Golden Flower: Chinese book of life* (London: Kegan, Paul: 1954). Translated from Chinese and explained by Richard Wilhelm with a foreword and commentary by C. G. Jung. Translated by Cary F. Baynes from the German edition of 1929.

Illustration Credits

Chapter 6 • Pauli, Heisenberg, and the Great Quantum Breakthrough

95 AIP Emilio Segré Visual Archives / Gift of Jost Lemmerich.

Chapter 7 • Mephistopheles

116 Courtesy Karl von Meyenn.
122 Niels Bohr Archive.

Chapter 8 • The Dark Hunting Ground of the Mind

126 FCGJ.

Chapter 9 • Mandalas

152 Courtesy of William Byers-Brown and Suzanne Gieser.

Chapter 10 • The Superior Man Sets His Life in Order

168 CERN Archive.

Chapter 15 • The Mysterious Number 137

257 University of Chicago, courtesy AIP Emilio Segré Visual Archives.
267 Courtesy Roy Glauber.

Notes

307 Courtesy Norman Ramsey.

Insert

SWCJ.
SWCJ.
SWCJ.
Photograph by Nordisk Pressefoto, Niels Bohr Institute, courtesy AIP Emilio Segré
    Visual Archives, Margrethe Bohr Collection.
CERN Archive.
CERN Archive.
CERN Archive.
CERN Archive.
CERN Archive.
CERN Archive.
CERN Archive.
CERN Archive.
CERN Archive.
CERN Archive.
Archiv der Max-Planck-Gesellschaft, Berlin-Dahlem.

Index